The Pocket Watch

The Pocket Watch

Restoration, maintenance and repair

Christopher S. Barrow

NAG Press

an imprint of Robert Hale · London

© *Christopher S. Barrow 2009*
First published in Great Britain 2009

ISBN 978-0-7198-0370-3

NAG Press
Clerkenwell House
Clerkenwell Green
London EC1R 0HT

NAG Press is an imprint of Robert Hale Limited

www.halebooks.com

A catalogue record for this book is available from the British Library

2 4 6 8 10 9 7 5 3 1

Printed in China and arranged by
New Era Printing Company Limited, Hong Kong

To Anne

Contents

List of illustrations

INTRODUCTION

This guide is intended to be a practical introduction to the restoration, maintenance and repair of mechanical pocket watches. It is written for those who have some basic knowledge of how a pocket watch works and want to extend this to include maintenance and repair.

It is based on my own experiences, gained over the last twenty years or more repairing and restoring a wide range of watches, dating from the eighteenth to the early twentieth century. There are a range of techniques that I have refined or developed over this period of time. This is not meant to be a definitive guide, and as you gain confidence I hope you will be able to develop and adapt your own methods.

The aim is not to return any watch to a 'perfect' condition, but to let it work reasonably well given its age.

Over the last 200 years or so of pocket watch manufacture, there have emerged five main types:

- The keyless going barrel lever (late nineteenth to twentieth century)
- The going barrel key-wound cylinders (nineteenth to early twentieth century)
- The English fusee full-plate lever (nineteenth century)
- The English three-quarter-plate centre seconds chronograph
- The English fusee verge (sixteenth to mid-nineteenth century)

There are a large number of other, more unusual, movements but most of these are variations on one or more of the above types and therefore, with a little adaptation, the same methods can be applied.

HOW A POCKET WATCH WORKS

A pocket watch is defined as a portable device for measuring and displaying the time of day, small enough to fit in a pocket. It does this by accurately controlling the release of stored energy. This energy is stored in the main spring, which is connected to a number of wheels (cogs), and a regulator, the escapement. Hands are attached to the ends of some of the wheels and these are used to indicate the time.

The main spring consists of a length of spring steel wound tightly inside a barrel (known as a spring barrel). The spring is attached at both ends, the outer to the inside of the spring barrel and the inner to an arbor at the centre. The spring barrel is connected (directly or indirectly) to a number of wheels held between plates, which form the watch train. These are used to transmit the energy stored in the spring to the escapement mechanism. The ratio of the number of teeth on each wheel and pinion determines the rate at which they rotate.

The most important part of a watch is the escapement. This provides the means of regulating the rate at which a watch runs. There are two main categories of watch escapement: the frictional rest and detached. In the former, the means of providing impulse to the balance is constantly in contact with the balance and hence can interfere with its movement. In the latter, the balance is only given impulse and hence in contact with it for a very small part of the total time the balance is in motion.

Types of escapement

All escapements are of the same basic construction – that is a balance wheel mounted on a staff, the ends of which are held in pivot holes. The balance is prevented from rotating beyond a certain limit by the hair spring. This arrangement means that, if free from external influences and given a flick, the balance will oscillate for quite a few seconds before stopping because of friction. In order to prevent this friction stopping the balance from oscillating, energy must regularly be imparted to it. The means of transferring this energy is what differentiates one type of escapement from another. The following list of escapements represents those most likely to be encountered:

- **Cylinder** – Frictional rest. Consists of a number of arrow-shaped 'teeth' mounted equidistant around the edge of the escape wheel. At any given time, one of these will be resting on the outer or inner surface of the partly cut-away cylinder which forms the balance staff. As the escape wheel teeth slide against the edge of the cut-away cylinder they push the balance one way, then the other, so keeping it swinging.

- **Verge** – Frictional rest. A crown wheel with an uneven number of teeth is mounted at right angles to the balance staff. The balance staff has two pallets set at approximately 100 degrees, one that engages with the top crown wheel teeth and one that engages with the bottom teeth. The force applied to the crown wheel causes first the top pallet and then the bottom pallet to be pushed away from the teeth, thereby releasing one tooth ready for the next one to engage with the other pallet. This causes the balance to swing first one way and then the other.

- **Table-roller lever** – Detached. The balance has a disc (table) attached to the balance staff. This has a small crescent cut out of it; also attached to the disc (in the middle of the crescent) is the impulse pin. Force is transferred to the balance via the lever which has a fork at one end, which engages with the impulse pin. The lever is moved from side to side by the escape wheel. This has a number of spur teeth which alternately 'lock' and 'unlock' against the pallets on the lever pushing it from side to side.

- **Spring detent** – Detached. A length of sprung steel is fixed at one end to a stiff piece of steel. The spring extends a small distance beyond the end and engages with a pallet on the balance staff. As the balance rotates one way, the pallet pushes the spring out of the way, and as it goes the other way, it pushes the spring and stiff bar. This disengages another pallet that holds the escape wheel, which as it rotates, impinges on the balance staff pallet and gives the balance an impulse. This is repeated for each forward rotation of the balance.

- **Rack lever** – Frictional rest. This works in the same way as the table roller lever as far as the escape wheel and lever are concerned. The other end of the lever has a segment of a cog (about a $^1/_6$) which engages with a rack on the balance staff. As the lever is made to move backwards and forwards by the escape wheel, it acts on the rack and so the balance is also swung backwards and forwards.

- **Massey types I–V** – Detached. This is very similar in design to the table roller lever and only really differs in the design of the impulse pin and lever fork. There are five variations on the design, numbered I, II, III, IV and V.

- **Savage two pin lever** – Detached. Again this is very similar to the table roller lever. The difference is that impulse to the balance is via a pin on the lever which engages with a slot cut into the table attached to the balance staff. There are two pins that are mounted either side of this slot, and these engage with the fork at the end of the lever and prevent it from accidentally disengaging due to a shock to the watch (known as the safety action).

- **Debaufre** – Frictional rest. Similar to the verge except that the crown wheel is replaced by a pair of star-shaped wheels, which act on two pallets on the balance staff, set at an angle to each other. As each star wheel is brought to bear on one of the pallets, the balance staff is pushed one way until the point passes the pallet, then the other one acts on the other pallet and forces the balance in the other direction.

- **Duplex** – Frictional rest. The escape wheel has two rows of teeth, one short and one long. The longer tooth rests on the balance staff as it rotates; the tooth eventually drops through a notch. As it does this, the small ring of teeth on the escape wheel engages with a pallet on the balance staff and gives it a 'kick'.

For animated views of some of the above escapements, take a look at The Horology Source, (see appendix A for details).

TOOLS

The following could be seen as an ideal list of tools, but it is perfectly possible to clean and carry out basic repairs, with just a few of them (those marked with an asterisk are the ones I consider to be essential). It is probably better only to buy the basic tools to start with, and then purchase the others as and when you need them. Most of the basic tools should be readily available (see supplier internet sites listed in appendix A), but the specialized ones may prove to be a little more difficult to obtain. It is always worth keeping an eye out on auction sites or at antique fairs for second-hand ones.

It is a good idea to buy the best-quality tools you can afford. Cheap ones will often perform less well and may even cause damage to the watch. One of the best makers of watch tools is Bergeon. This Swiss company produces a huge range of tools and materials specially designed for the watch repairer. They are not cheap, but they will repay the initial cost by giving years of good and reliable service.

Arkansas stone
Brass cleaner *
Burnisher
Case opener *
Craft knife with retractable blade *
Cutters (strong enough to cut steel) *
Diamontine polishing powder
Fibber glass cleaner
Light source (good quality and adjustable) *
Glue *

Hammer (brass faced)
Hands remover *
Heat source (spirit lamp or gas stove)
Jacot tool
Jeweller's hammer *
Two pairs of high-quality tweezers (anti-magnetic) *
Magnifying glass and eye pieces (if you wear glasses, consider
  a clip-on one) x3 and x10 *
Needle files *
Oil storage box and watch oil *
Pair of flat brass-faced pliers *
Pair of flat-faced pliers *
Peg wood *
Pin vice *
Punches (round ended)
Reamers (cutting broaches)
Rubber work mat
Screwdrivers (0.8mm to 4mm), ideally with replaceable
  blades. For ease of use, it might be worth looking for a set
  which is stored in a rotating storage base *
Screwdriver sharpening tool used to hold the screwdriver at a
  fixed angle to the stone.
Segmented storage tray with lid *
Sharpening stone (diamond impregnated type)
Silver cleaner *
Staking tool
Steel wool (000 grade)
Taps
Watch keys (Nos 0–12) *
Wire in a range of sizes (brass and steel)

PREPARATION

I t is important to arrange your work place before starting on any watch. Make sure that you are seated at a comfortable level to the desk or bench and that you have enough free space to work in. Pay particular attention to the lighting: a good, moveable lamp is very useful. Make sure the work area is clean and free from dust. Use a piece of white paper (A3) to work on; this makes it easier to see the parts, and it should be replaced when it becomes dirty. Keep a few layers of paper underneath it to protect the desk surface. Better still, use a specially designed rubber mat, which will help to prevent parts from rolling off the desk.

If you are not very familiar with the workings of a pocket watch, it might be worth taking a few photographs or making sketches during the dismantling process. You can then use these to determine which parts fit where as you reassemble them.

Oil is normally supplied in small bottles; it is good practice not to use it straight from the bottle, as there is a chance that it will become contaminated. It is much better to buy an oil-storage box. This is a small container about 30–40mm in diameter with a depression in the centre and a tight-fitting lid. To use it, transfer a small amount of oil into the depression, and when this has been used or becomes dirty, clean it out and refill it. This way you are only using clean, fresh oil on your watches.

It is very important to use the right-sized screwdriver. This means the blade width should be the same as or just less than the width of the screw head. If it is wider, there is a risk of it catching on the surrounding metal and damaging it; if it is smaller, there is a risk of damaging the screw slot. The blade should be the correct thickness so that it forms a tight fit in the slot.

This is probably a good time to mention a particular problem when working on watches, which is that there are quite a few parts whose sole purpose seems to be to fly off at an odd angle and hide themselves (I am convinced that some of them grow legs as they fly through the air, so that when they land on the floor, they can run under an item of furniture and hide). This means hours spent on your hands and knees trying to find some vital part. There is no easy way to avoid this problem other than to be aware that, as you dismantle a watch, some parts are likely to fly off if you are not careful.

A word of warning: when you obtain a watch, even if it looks to be in good condition, do not be tempted to wind it up until you have dismantled, cleaned and oiled it. There is a very real chance that parts will have seized up owing to the old oil drying out; this is especially true of the main spring, which may break if put under strain after being allowed to dry out.

WHERE TO OBTAIN WATCHES

Over the years, I have been able to obtain suitable watches for repair from a number of sources:

- Antique shops (you may have to ask whether they have any broken watches as these are not normally on display)
- Second-hand, junk and charity shops
- Auctions (online and traditional)
- Antique and collectors' fairs
- Your own family and friends

If you obtain watches from friends or family, make sure that you are either fully capable of repairing them or that they accept the consequences of the watch being returned in a less than perfect condition! It is probably better to practise on watches from other sources before tackling those of friends and family.

THE KEYLESS GOING BARREL LEVER

I t is best to start with low-value watches until you are confident enough to tackle more expensive ones. The best watch to consider first is probably the keyless going barrel lever. Most examples will be either Swiss or American, with a few English ones, dating from the late nineteenth and early twentieth centuries. They usually come in silver, nickel or gold-filled (plated) and (much more rarely) gold cases, often plain, but sometimes with engine turning or an embossed pattern.

In general they come in three forms: open-faced (figure 1), half-hunter and full hunter. As far as repairing is concerned, there is not a great deal of difference between the movements used, so, with only slight variations, the following procedures can be applied to all three types.

Figure 1 Open-faced keyless going barrel lever

Removing the movement from the case

Remove the front bezel using a case-opening knife, which has a blunt blade and is specially designed not to damage the case; DO NOT use a penknife or other shape of blade, as this will risk damaging the case. Be aware that on some open-faced watches, the bezel, and sometimes the back, is screwed on. If this is the case, hold the bezel with your thumb and forefinger and turn it anticlockwise. Using rubber gloves or a piece of rubber may give a better grip. If it proves to be very stiff, apply a small amount of oil around the junction between the case and the bezel or back, and allow it to seep in overnight. Sometimes the bezel can be 'cross threaded'. The only way to remedy this is to use a case opener and force it onto the correct threads. Try to use as little force as possible, as this may damage the bezel or the case.

Figure 2 Using a case knife to remove the front bezel

Figure 3 Removing hands

Once you have removed the front bezel, remove the hour and minute hands using a hand remover. This is designed to lever the hands up without damaging the dial. Pressing the button on top of the hand remover will cause the jaws to open. Press the tool down over the hands (the centre section is spring loaded) and release the button. The jaws should close under the rim of the hour hand. Pushing the whole handle down should cause the jaws to lever up the hands and free them from the cannon pinion. Be careful not to bend the hands; position the hand remover so that it does not put any strain on them. The second hand can be carefully removed by holding a craft knife near the stem and levering it up; hold your finger over it so that it does not fly off.

Figure 4 Rear covers

Turn the watch over, and open the back in the same way as the front bezel: either unscrew or use a case opener.

Watches with screw-on backs do not usually have an inner cover, but ones with hinged backs often do, and this will also need to be opened with the case opener.

Some people use their nails to open watch cases; this will not damage the watch, but can be very painful on the nails, and is therefore not to be recommended.

Letting down the main spring

Before removing the movement, it is *very* important that you release the tension in the main spring. In order to do this you need locate the 'click' stop that engages with the winding wheel (the large, usually chrome-finished wheel that is attached to the spring barrel, near the winder – see figure 13). Rotate the winder a little clockwise (the normal direction used to wind the watch); you should see the click move away from the wheel. Still keeping pressure on the winder, carefully hold the click away from the wheel and then allow the winder to rotate anticlockwise between your fingers (do not let go as this could damage the spring). You may need to allow the click to re-engage with the wheel from time to time, so you can keep hold of the winder as it rotates while the spring is wound down. Make sure that there is no tension left in the spring before you proceed to the next step. Do not use your fingers to hold the click away from the wheel as you can do quite a lot of damage if the spring runs down too quickly!

There are two types of winder fitted on going barrel levers. One is an integral part of the case (more usual on American watches); the other passes through the case and is held in place by a screw and bar on the movement. For the former, you only need to remove the two screws at the edge of the movement (see figure 5). These are to be found at roughly 4 o'clock and 10 o'clock as viewed

These screws hold the movement in place

Figure 5 Position of movement-retaining screws

from the rear, and overlap onto the edge of the case. Once they have been removed, the movement can be pushed carefully from the back near the bottom. It may be necessary to pull the winding crown out before the movement can be removed completely.

If the winder passes through the case, then you first need to loosen the screw holding it in place (two or three turns of the screw). The screw is usually located close to the winder stem. Once it has been loosened, the winding crown and stem should pull free of the movement and case. To complete the job, you need to remove the screws that are holding the movement in place. There may only be one (near the bottom).

Sometimes these screws are partly cut away and only need to be turned through 180 degrees or so to release the movement.

Removing the dial

Most dials on good-quality watches are made of enamel, consisting of a layer of ground glass mixed with colouring mate-

Dial-retaining screw

Figure 6 Position of dial-retaining screw

rial (usually white), which is then fused onto a copper disc at high temperature. This produces a very hard, smooth surface which will not fade or discolour. The numerals are then painted on and the dial fired again to fuse them into the enamel. The only drawback with this is that the dial is prone to damage if the watch is dropped or mishandled (producing either chips or hairline cracks).

Most dials are held in place by either screws at the side of the movement or a specially cut screw on the back of the dial plate. If it is held by side screws, loosen all of them with a suitably sized screwdriver (there is no need to remove them completely). If it is held by specially shaped screws with a semicircle cut out

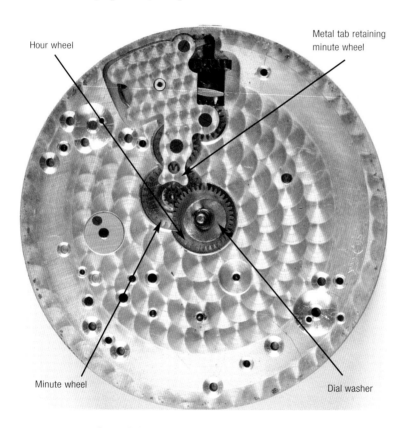

Figure 7 Under-dial view showing motion works

(see figure 6), turn the screw anticlockwise with a screwdriver until it disengages from the dial foot. Then, using a flat blade, carefully lift the dial away from the movement. Try not to flex the dial too much as this may cause it to crack.

With tweezers, remove the motion work and, if there is one, the dial washer (sometimes only the hour wheel can be removed as the minute wheel may be held in place by a metal tab).

As each part is removed, place it in a storage tray. The best types have divisions to keep the parts separate and a clear domed cover to protect them until the watch is reassembled.

Cleaning the dial

If the dial is undamaged, it only needs to be cleaned in soap and warm water (not too hot, otherwise the enamel might be damaged). After cleaning, dry it off with a paper towel. If the dial has hairline cracks, these can sometimes be made less obvious by soaking the dial in a mild bathroom cleaner (Bath Power by OzKleen seems to work well). It is normal for the numbers to be an integral part of the enamel dial and they are therefore unlikely to be removed by any cleaning. However if there is other writing on the dial (maker's name etc.) care must be taken to ensure that it is not removed during cleaning. Check a small area first.

Once cleaned, put the dial safely to one side.

Removing the cannon pinion

After the motion work has been removed, the cannon pinion is next. This is held in place by a friction fit. To remove it, you need to take hold of it with a pair of pliers, and with a twisting action pull it firmly away from the front of the watch. Be careful not to apply too much force as you may distort the cannon pinion. If it proves to be very stiff, a small amount of oil should be applied and left overnight to work its way in.

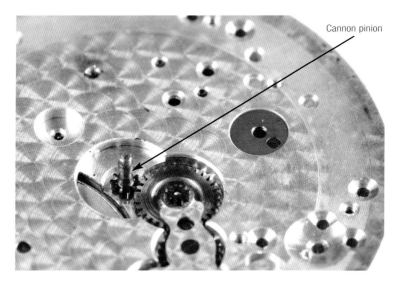

Cannon pinion

Figure 8 Cannon pinion

Removing the balance

The balance is held in place by the balance cock, which in turn is fixed in place by a single screw. Hold the movement face down and level, then undo the screw and remove it. Using a strong pair

Figure 9
Removing balance
cock

of tweezers carefully lift up the balance cock. Note that it is also kept in place by pins that engage in holes on the main plate. These can be quite a tight fit and so you may need to lever the cock away from the main plate. There may be a small slot cut into the cock where it meets the main plate (see figure 9). This can be used to help lift the cock free of the plate. Once free, carefully lift it away, making sure that the balance is free, and then lift the complete assembly clear of the movement. Turn the balance cock over so that the balance is resting on it. Be very careful not to stretch the hair spring.

Curb pins, below hair spring

Figure 10 Balance assembly

To separate the balance from the balance cock, locate the small screw on the side of the balance cock, where the outer end of the hair spring is fixed. Loosen this and lift the hair spring stud away. Be careful as it may still be held by the curb pins of the regulator. Using a pair of tweezers held either side of the curb pins, carefully lift the hair spring clear. The balance should now be free of the balance cock.

Removing the lever

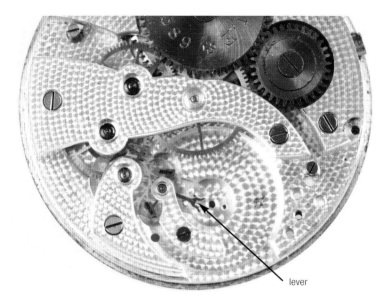

lever

Figure 11 Position of lever

The lever is held in place by either a bridge held on both sides by screws or a curved bridge with a single screw. Remove the screw(s) and carefully lift the bridge away from the plate using tweezers (be very careful to lift it straight up or you might bend or break the lever pivot). Then lift the lever away and place it with the bridge and screw(s) in the storage tray.

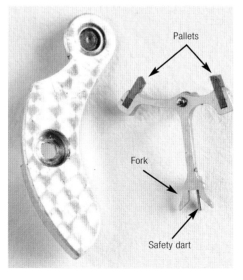

Pallets

Fork

Safety dart

Figure 12 Lever and mounting bridge

35

Removing the wheels

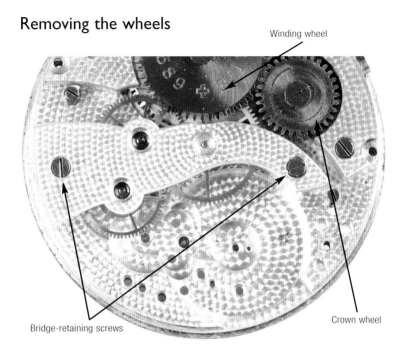

Winding wheel

Crown wheel

Bridge-retaining screws

Figure 13 Removing winding and crown wheels

Depending on the type of watch movement, the wheels (centre, third, fourth and escape) are held in place by removable bridge(s). These in turn are held in place by screws. Before the wheels can be removed, the winding and crown wheels need to be removed. These are the smaller and larger, usually chrome-finished wheels that connect the winder to the spring barrel. To remove the winding wheel, use a suitable screwdriver and remove the screw. Lift the wheel up using tweezers and place it with the screw in the storage tray.

The crown wheel is also held in place by a screw, but care must be taken when removing this as it sometimes has a left-handed thread. Use only light pressure until you are sure which way it unscrews. Place the crown wheel in the storage tray with its screw; if it is left-handed, do not mix it up with any other screws.

Once these two wheels have been removed, you can remove the train wheels. If there is only one plate holding the wheels in

Bridge-retaining screws

Figure 14 Removing bridges

Third wheel

Centre wheel

Fourth wheel

Figure 15 Position of wheels

place remove all the screws (usually three) and lift the plate free; you will now be able to remove each wheel separately. Note that sometimes the screws may be of different lengths, so it is very important that when you reassemble the watch, the correct-length screw is returned to the same location. If the wheels are held in place by separate plates, remove each one and then remove the associated wheel. Make sure that you know the order in which they were removed as they are not interchangeable. When lifting any wheel away from the plate, always lift it perpendicular to the plate so as to avoid bending or breaking the pivot.

Removing the spring barrel

There should only be one plate still screwed into the main plate; this holds the spring barrel. Remove the screws that hold it in place and lift the plate free. You should now be able to remove

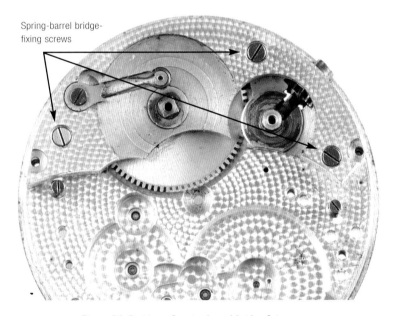

Spring-barrel bridge-
fixing screws

Figure 16 Position of spring-barrel bridge-fixing screws

the spring barrel. Depending on the design of the winder (fitted to the case or held in place by a screw), you will see either two separate cogs which can be removed by tweezers or an assembly which is held in place by other parts. This section of the watch is one of the more difficult to deal with, so pay particular attention to the position of the various parts.

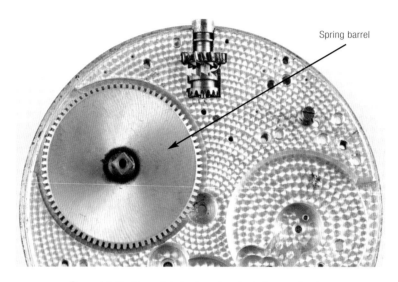

Spring barrel

Figure 17 Position of spring barrel and winding mechanism

Checking the spring

One of the more common causes of a watch not running is a broken spring. To check this, remove the spring barrel cover by inserting a screwdriver into the slot and carefully levering up the cover. Check that the spring is working and that it is still connected at both ends. Either the outer end should have a hole which hooks over a pin on the inside of the spring barrel or the end of the spring should be bent back to catch on a notch on the inside of the barrel. The inner end should fix onto the hook on the barrel arbor. If this all looks all right, then you only need to apply a small amount of oil and replace the cover. This fits into

Figure 18 Opening the spring barrel

Figure 19 Inside the spring barrel

a small groove and may need to be pressed back into place with a pair of brass-faced pliers. It should be flush with the barrel when properly fitted.

One variation on the way the outer end of the spring is held is two small pins sticking out at right angles to the spring, which engage in two holes, one in the bottom of the barrel and one in the cover. It is therefore necessary to make sure that the hole in the cover is lined up with the pin before it is clicked back into place.

Waltham watches, and some other American ones, use a slightly different form of construction for the spring barrel. This consists of a wheel with an integral hook for the inner end of the spring and a separate spring barrel where the outer end of the spring is connected. The arbor is a separate part that passes through the wheel and has a small square section that engages with the spring barrel. This arrangement may feel a little odd, as though it is broken. In order to oil this type, carefully separate the two parts, making sure that you do not pull the spring out of the barrel, and apply oil to the spring. Push the two halves together and refit the arbor. Take care when fitting this type into the plates; make sure that the two parts are correctly aligned before screwing the plate in place.

Cleaning the main plate and bridges

The best way I have found to clean most watch parts is to use a metal polish. Some older books on watch repairing advocate the use of benzine, but this is not only difficult to obtain but also a poison and should therefore not be used. Using kitchen towel and metal polish, rub the surface until all the dirt, marks, grease and old oil have been removed. After polishing them with a clean paper towel, the next step is to clean the pivot holes. This is done with a length of peg wood (see list of suppliers in appendix A), whose end has been sharpened to a point with a craft knife. Insert this into the pivot hole and rotate it a few times, repeating until the wood comes out clean. Do not be tempted to reuse the peg wood on another pivot hole without resharpening it, as this can lead to the point breaking off and sticking in the hole. If this happens, resharpen the end and try to push the broken bit out from the opposite side. When you are happy that all the pivot holes are clean, return the parts to the tray.

If the pivot holes are straight into the metal (i.e. not jewelled), check that they are still round by examining them with a magnifying glass. If they are not, it may be necessary to close up the hole with a staking tool or, if they are very worn, rebush them.

To close the hole, place the plate on the staking tool with the inside surface facing up, and with a round-ended stake lightly tap around the pivot hole (being careful not to hit the hole itself as this may make it larger) until it is a little smaller than required for the pivot. Then rotate a suitably sized burnisher inside the pivot hole until the pivot just fits without any friction. This is something that can really only be judged by trial and error, so take it slowly. If the hole is too large to be closed up by staking, a new bush may need to be inserted. This consists of a small, preformed brass disc with a hole in the middle. The old pivot hole must be drilled out (using a drill a little smaller than the diameter of the bush). Then, using a reamer, working from the inside, open the hole until the bush is just too big for the hole. The reamer should have produced a slightly conical hole in the plate. The bush can then be fitted in place by positioning it over the hole (from the inside) and, using a flat-faced stake, tapped into place. Once you are happy that it is correctly positioned, open up the hole with a burnisher as above.

Cleaning the wheels and pinions

Examine each wheel to make sure that it is flat and that the pivots are straight. If so, clean both pivots by holding a piece of peg wood so that the pivot sinks into the blunt end and rotate it a few times. If the pinion vanes have dirt between them, this needs to be cleared out. Sharpen a piece of peg wood to a flat point and push along each of the vanes until all the dirt has been removed. A rub with an old (clean) toothbrush can remove most general dirt. Do not use too much force as this can distort the wheel.

Reassembling the watch

Assuming that there are no parts missing and any necessary repairs have been made, it is now time to reassemble the watch. First put the wheels in place; if each is held by a separate plate, fit them one at a time. You may need to check the order in which they are put back. Start with the one whose wheel is nearest the main plate (this is when you might need to refer to any photographs you took at the start). If the wheels are all held by one plate, then you will need to fit all of them before the plate can be put in place. Hold this lightly in position (do not try to screw it down yet), and check that all the pivots pass through the appropriate pivot holes (top and bottom). You may need to use a pair of tweezers to move the wheels until the pivots have engaged with their pivot holes. When you are happy that all the pivots are in place, apply a little pressure to the centre wheel; if everything is correctly in place, all the wheels should rotate freely. Then, and only then, screw in each screw. Do not tighten them fully at this stage but stop as soon as you can feel some resistance. Make sure that all the wheels can still turn freely, and then tighten the screws.

Assemble the winding mechanism. Then, holding the main plate face down, fit the spring barrel in place. Fit the spring-barrel bridge and make sure that it is flush with the main plate. When you are happy that it is in place, fit the screws and screw them home. Never overtighten them, and always use the correct size of blade otherwise you may damage the head. Light pressure applied to the spring barrel should cause the complete train to move freely. If not, you may need to loosen the screws and check that the barrel is correctly positioned.

Now fit the lever carefully the correct way up into its bottom pivot. Then place the bridge over it so that the top pivot engages with the pivot hole. Insert the screw(s) and lightly tighten them. Make sure that both the bottom and top pivots are correctly engaged (check with a magnifying glass) then, when you are happy that the lever can move freely, tighten the screws.

Before fitting the balance, lightly oil the pivots, including the bottom balance pivot. Do not overoil, as this can lead to a build-up of dust. Use an oil pin (a tool specially designed for the job) and transfer a drop to each pivot; aim to put on just enough to partially fill the indentation around the pivot. Also apply a small amount of oil to the lever pallets.

Now carefully lift the balance and balance cock and turn it over so that the balance is dangling down (be careful not to let the balance catch on anything or be shaken too much as this could distort the spring). Place the bottom pivot into the lower pivot jewel (which means that the balance needs to be put under the centre wheel). Then line the balance cock up so that the mounting lugs on the bottom drop into the holes in the main plate. Keeping the balance level with the main plate, very carefully push the balance cock down until it is flush with the main plate. Make sure that, as the balance is pushed down, the top pivot engages with the jewel in the balance cock. If the balance is correctly positioned, it should swing freely when lightly shaken. If it moves but appears to come to a stop against something, then the problem is most likely that the impulse jewel is not engaged correctly with the lever. For the balance to swing freely, the impulse jewel must pass through the notch on the lever as it passes. If this is the case, you will need to lift the balance cock and twist the balance to the right or left before repeating the above steps to fit the balance cock. When you are happy that it is correctly positioned, you can screw it in place. As you tighten the screw, it is a good idea to give the balance a light shake so that it is oscillating to make sure that everything is in the correct position as the screw is tightened.

At this point, if all has gone well, you should have a working watch. Using the winder, wind the watch up (at this stage only turn it a few times). Then shake it and it should start ticking! If it does, then wind it fully. Always be cautious when winding up a watch, especially one that has just been cleaned. You should feel the spring become stiffer as it approaches the end; it is always better to stop before it is fully wound, to prevent damage to the spring or another part of the watch. Turn the movement over and make sure that it still ticks. Place it somewhere safe and check it after twenty-four hours. If it is still running, then you can fit the dial.

Fitting the dial

Fit the cannon pinion and push it fully home. If the minute motion work is in place, check carefully that the teeth of the cannon pinion engage with it, before you push it home. Fit the motion work and dial washer. Fit the dial and, depending on the fixing method, either tighten the side screws or rotate the cutaway screws so that they engage with the dial feet.

Fitting the hands

If the hands are of blue steel, they only need to be carefully wiped and fitted. If they are gold, they may need to be cleaned with metal polish. Lay them on a flat surface and wipe along their whole length with a piece of kitchen towel soaked in polish. Do not wipe towards the centre mounting as it is very easy to catch the point and bend the hand. Finish by cleaning with a fresh piece of kitchen towel.

Once clean, fit the second hand. Press it down with the back of a pair of tweezers until it is almost flush with the dial, making sure it is not in contact at any point. If it is, this may cause the watch to stop and can also leave marks on the dial. Next fit the hour hand, pressing it down with a pair of tweezers. Move it, using the winder (never try to push hands as this will cause them to bend or brake), so that it is pointing to 12 o'clock, then fit the minute hand so that it is in line with the hour hand and press it into place with tweezers. Using the winder, rotate the hands through twelve hours to make sure they do not catch (always move them in a clockwise direction to prevent possible damage to the movement).

Regulation

Once the watch is working, you will need to bring it to 'regulation'. This involves adjusting the regulator arm on the balance cock until the watch keeps good time. The regulator arm usually moves over a scale engraved on the balance cock, usually marked F (fast) and S (slow) but sometimes R and A (retard and advance). As the arm is moved, it moves the curb pins along the hair spring, so increasing or decreasing its effective length. The effect of changing the length of the hair spring is to change the rate of rotation of the balance (the beat) – the longer the hair spring, the slower the beat, the shorter the spring the faster the beat.

Things that might stop the watch from working

Like all mechanical devices, watches do not always work as intended. This may be due to a number of reasons. It may be a damaged or badly fitted part; there may be too much friction in the train or any number of other faults that prevent it from working properly. If the balance will not swing or moves sluggishly, the problem might be that the safety pin is rubbing on the edge of the table roller. If this is so, remove the lever, and holding it in a small vice carefully bend the safety pin back a fraction.

Another problem is a lack of power to the lever. Remove the balance and balance cock then, using a piece of peg wood, move the lever from side to side. If it is moved a small amount, it should snap back when you let go. If you move it a bit more, it should snap over to the other side. If this does not happen, the problem lies in the train. Let down the tension in the spring and remove the lever, then wind it up a little; the escape wheel should rotate freely. Do not wind it up too much, as it could damage the wheel pinions if they are allowed to run too fast. If it does not turn freely, check that there are no bits trapped in the leaves of the pinions or between the teeth on the wheels.

Another cause of a sluggish train may be a distorted fixing plate. If you slacken off the tension on the screws that hold the plates in place, and the train becomes free, then the plate might be bent. Dismantle it and check it against a flat edge. You can straighten them by applying a small amount of pressure in the opposite direction. Reassemble and check if the train is running free when the screws are tight.

Fly-back chronograph

I mention this type of watch here, as it is basically a variation of the standard keyless going barrel discussed above. If you open the back of one of these watches, you may not at first recognize that it is similar in construction. Closer examination, however, will show that this is in fact a standard going barrel with additional parts added to provide the fly-back chronograph function. Before attempting any work on one of these watches, I would suggest that you gain experience of the simpler type of movement first.

If you feel confident to proceed, it is imperative that you either take photographs or make sketches of the movement before you remove any parts. Most of the additional parts are usually attached to the upper surface of the bottom plate covering the spring and wheels. If the chronograph mechanism is clean and functioning correctly, then it may not be necessary to dismantle it. If this is the case, then following the above process should lead to a working watch.

If the chronograph function is not working, then you will need to disassemble this part of the watch as well. Follow a similar process to that outlined above, and take photographs as this part of the watch is very complicated.

THE GOING BARREL CYLINDER

This type of watch is based on the cylinder escapement invented by George Graham in the 1720s. It was not taken up by the watch trade in any great numbers, however, until the start of the nineteenth century, when it became very popular in France and Switzerland. The going barrel cylinder that I describe below became common from the mid-nineteenth century onwards.

Like the going barrel lever, this kind of movement can also be found in a wide range of case styles, from small decorative ladies' fob watches to oversized Goliath pocket watches. Again as with the going barrel lever, as far as repairing these types of watches are concerned, there is not a great deal of difference between the movements used so, with only slight variations, the following description can be applied to most types.

I describe here a typical French-made going barrel cylinder pocket watch. Apart from scale, the description should be applicable to most watches you are likely to come across. Earlier examples (pre-1860s) are constructed along the same lines as the verge and fusee lever, and should therefore be cleaned and serviced in a similar manner, as described later.

Removing the movement from the case

Remove the front bezel using a case-opening knife, which has a blunt blade, and is specially designed so as not to damage the case; *do not* use a pen knife or other blade as this will damage

Figure 20 Open-faced going barrel cylinder watch

Figure 21 Inner and outer back covers

the case. Be aware that on some open-faced watches, the bezel, and sometimes the back, is screwed on. If this is the case, hold the bezel with your thumb and forefinger and turn it anticlockwise. Using rubber gloves or a piece of rubber may give a better grip. If it proves to be very stiff, apply a small amount of oil around the junction between the case and the bezel or back, and allow it to seep in overnight. Sometimes the bezel can be cross threaded. The only way to remedy this is to use a case opener and force it onto the correct threads. Try to use as little force as possible, as this may damage the bezel or the case.

Sometimes the front bezel may be hinged, in which case it only needs to be opened (do not try to open it too far, as this may damage the hinges).

Turn the watch over, and open the back in the same way; either unscrew or use a case opener.

Letting down the main spring

Before removing the movement it is *very* important that you release the tension in the main spring. If you find that there is no tension in the spring or it will not wind, this may be due to a broken main spring. If this is the case, you can skip the next section.

Key-wound

In most cases, the spring barrel is suspended from a bar held in place by a screw on either side. It is wound by a key which acts against the ratchet. To let down the spring, fit a correctly sized key on the winding square and turn it a little in the direction of winding (clockwise). This should push the spring that engages with the ratchet away from it. Use a small screwdriver to keep the spring away from the ratchet, and slowly allow the key to turn between your fingers. You may need to allow the spring to re-engage with the ratchet from time to time so that you can reposition your fingers on the key.

Keyless

You need to locate the click stop that engages with the winding wheel (the large, usually chrome-finished wheel that is attached to the spring barrel, near the winder). Rotate the winder a little clockwise (the normal direction used to wind the watch). You should see the click move away from the wheel. Still keeping pressure on the winder, carefully hold the click away from the wheel and then allow the winder to rotate anticlockwise between your fingers (do not let go as this could damage the spring etc.). You may need to allow the click to re-engage with the wheel from time to time so that you can keep hold of the winder as it rotates while the spring is wound down. Make sure that there is no tension left in the spring before you proceed to the next step. Do not use your fingers to hold the click away from the wheel as you can do quite a lot of damage if the spring runs down too quickly!

Removing the winding crown

The winding crown and stem pass through the case and are held in place by a screw and bar on the movement. First you need to loosen the screw holding the winder in place (two or three turns); it is usually located close to the winder stem. Once it has been loosened, the winding crown and stem should pull free of the movement and case. To complete the job, you need to remove the screws that are holding the movement in place. There may only be one (near the bottom).

Sometimes these screws have part of the head cut away and only need to be turned through 180 degrees or so to release the movement.

Removing the hands

The hour and minute hands are removed using a purpose-made hand remover. This is designed to lever the hands up without damaging the dial. Press the button on top of the tool, which will

Figure 22 Removing hands

cause the jaws to open; press it down over the hands (the centre section is spring loaded) and release the button. The jaws should close under the rims of the hands. Press the whole handle down, which should cause the jaws to lever up the hands and free them from the cannon pinion. Position the tool so that it does not put any strain on the hands. The second hand can be carefully removed by holding a craft knife near the stem and levering it up; hold your finger over it so that it does not fly off.

Removing the dial

The most common method of retaining dials on cylinder watches is by a specially cut screw (with a semicircle cut out) on the back of the dial plate. Using a screwdriver, turn the screw anticlockwise until it disengages from the dial foot. Then using a flat blade, carefully lift the dial away from the movement. Try not to cause the dial to flex in any way, as it may crack. Be aware that sometimes the dial feet may have become bent. If this is the case, you may need to lift the dial on one side first before you can remove it completely. Be very careful about bending the dial feet, as this may cause damage to the front of the dial enamel. It is much better to leave them as they are.

Figure 23
Rear view of
movement

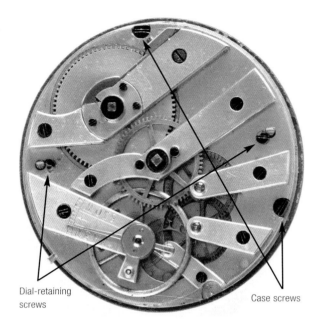

Dial-retaining
screws

Case screws

Figure 24
Under-dial
view of move-
ment

Hour wheel

minute wheel

With tweezers, remove the motion work and, if present, the dial washer (sometimes only the hour wheel can be removed, as the minute wheel may be held in place by a metal tab).

Cleaning the dial

If the dial is undamaged, it only needs to be cleaned in soap and warm water (not too hot, otherwise the enamel may be damaged). After cleaning, dry it off with a paper towel. If the dial has hairline cracks, these can sometimes be made to look less obvious by soaking in a mild bathroom cleaner (Bath Power by OzKleen seems to work well). It is normal for the numbers to be an integral part of the enamel dial and they are therefore unlikely to be removed by any cleaning. However if there is other writing on the dial (maker's name etc.) care must be taken to ensure that it is not removed during cleaning, so check a small area first. Once cleaned, put the dial safely to one side.

Removing the balance

The balance is held in place by the balance cock, which in turn is fixed in place by a single screw. Hold the movement face down and level, then undo the screw and remove it. Using a strong pair of tweezers carefully lift up the balance cock. Note that the cock is also retained in place by pins that engage in holes on the main plate. These can be quite a tight fit, so you may need to lever the cock away from the main plate. There may be a small slot cut into the cock where it meets the main plate. This can be used to help lift the cock free of the plate. Once it is free, carefully lift it away. Make sure that the balance is free paying particular attention to the teeth on the escape wheel as these can catch inside the cylinder, and then lift the complete assembly clear of the movement. Turn the balance cock over so that the balance is resting on the cock. Be very careful not to stretch the hair spring.

It is usual for the hair spring to be attached to the balance cock by a brass pin. If this is the case, push it clear of the hole with tweezers (be very careful not to damage the hair spring). The

*Figure 25
Rear view
showing
balance-cock
retaining screw*

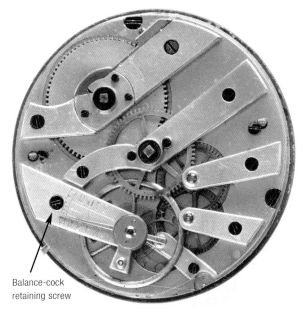

Balance-cock
retaining screw

Figure 26 Balance assembly

regulator will either be two curb pins or a pin and shaped cover. If it is the former, hold a pair of tweezers either side of the curb pins and lift it clear. If it is the latter, there should be a slot in the keeper. This should be turned through 90 degrees. The hair spring and balance can be lifted clear and placed in the storage tray.

On some cylinder watches the arrangement is similar to the going barrel lever, in which case proceed as follows. To separate the balance from the balance cock, locate the small screw on the side of the balance cock, where the outer end of the hair spring is fixed. Loosen this and lift the hair spring stud away; be careful as it may still be held by the curb pins of the regulator. Holding a pair of tweezers either side of the curb pins, carefully lift the hair spring clear. The balance should now be free of the balance cock. Place both parts in the storage tray.

Removing the cannon pinion

This consists of a setting pin with a square head that passes through the centre wheel and is very slightly tapered so that it fits tightly when fully home in the centre wheel. The cannon pinion then fits

Cannon pinion

Figure 27
Cannon pinion

on the other end, also with a friction fit. The end of the setting pin should stick out of the cannon pinion a small amount. To remove it, hold the movement and lightly tap the protruding end of the winding stem; it should come loose and the cannon pinion can be lifted clear. The setting stem can then be removed from the rear.

In keyless watches, the arrangement is similar except that the cannon pinion is fitted onto a pin that passes through the centre wheel (it looks a bit like a blunt dressmaker's pin).

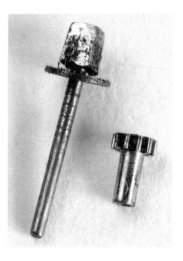

Figure 28 Hand setting arbor and cannon pinion

Removing the escape wheel

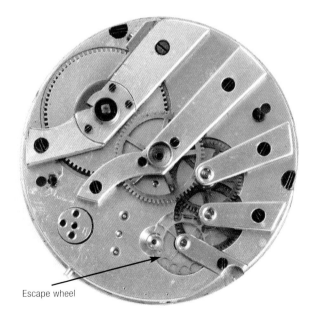

Escape wheel

Figure 29 Cylinder movement escape wheel

The escape wheel is held in place by a bar with a single screw. Remove the screw and carefully lift the bar from the plate using tweezers (be very careful to lift it straight up or you might bend or break the escape pivot). Then lift the escape wheel away and place it with the bar and screw in the storage tray.

Removing the wheels

Depending on the type of watch movement, the wheels (centre, third and fourth) are held in place by some form of removable bar. These in turn are held in place by screws. If the movement has a key-wound suspended spring barrel, you can remove all

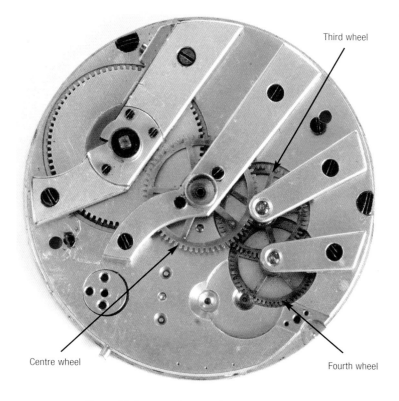

Figure 30 Position of centre, third and fourth wheels

the wheels by unscrewing the retaining screw(s) and lifting each bar clear. Make sure you keep the correct bar with its wheel.

If the movement looks like a going barrel lever, proceed as for that type of watch. Note that the screws may be of different sizes, so it is very important that when you reassemble the watch, the correct sized screw is returned to the same location. When lifting any wheel away from the plate, always lift it perpendicular to the plate so as to avoid bending or breaking the pivot.

Removing the spring barrel

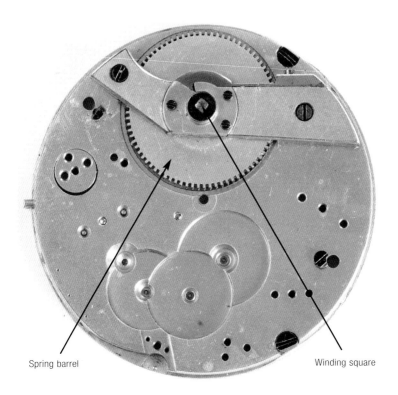

Spring barrel

Winding square

Figure 31 Position of spring barrel

There should only be one plate still screwed into the main plate, which holds the spring barrel. Remove the screws (usually two) that hold this in place and lift the plate free. The spring barrel should come away with the bar.

Checking the spring

One of the more common problems with pocket watches is that the spring can often break. To check the spring remove the spring barrel cover by inserting a screwdriver into the slot and

Removing spring barrel cover; note that this watch has had its Geneva stop work removed in the past, a common situation.

Figure 32 Removing spring barrel cover

carefully levering up the cover. Note that on some watches of this type, there is what is called Geneva stop work. This consists of a cross screwed to the cover and a shaped piece attached to the winding arbor, and is designed to control which part of the spring is used. In order to get into the spring barrel, you need to remove the shaped piece of metal from the arbor. This is held in place by a friction fit; use a screwdriver (not one of the good ones used to disassemble and reassemble the watch) and lever the stop piece up. You can now remove the barrel cover by putting the blade of a screwdriver into the slot and levering the lid off. Check that the spring is unbroken and is still connected at both ends. Either the outer end should have a hole which hooks over a pin on the inside of the spring barrel or it should be bent back to catch on a notch on the inside of the barrel. The inner end should fix onto the hook on the barrel arbor. If this all look all right, then you only need to apply a small amount of oil and replace the cover. This fits into a small groove and may need to be pressed back into place with a pair of pliers. It should be flush with the barrel when properly fitted.

One variation in the way the outer end of the spring is held is two small pins sticking out at right angles to the spring, which engage in two holes, one in the barrel and one in the cover. It is

Figure 33 Disassembled spring barrel

therefore necessary to make sure that the hole in the cover is lined up with the pin before it is clicked back into place.

Cleaning the main plate and bridges

The best way I have found to clean these parts is to use a metal polish. Some older books on watch repairing advocate the use of benzine, but this is not only difficult to obtain but is also a poison and should not be used. Using kitchen towel and metal polish, rub the surface until all the dirt, marks and grease have been removed. After polishing them with a clean towel, the next step is to clean the pivot holes. This is done using a length of peg wood (see list of suppliers in appendix A), whose end has been sharpened to a point with a craft knife. Insert this into the pivot hole and rotate it a few times; repeat until the wood comes out clean. Do not be tempted to reuse the peg wood on another pivot hole without resharpening it, as this can lead to the point breaking

Figure 34 Cleaning pivot holes

off and sticking in the hole. If this happens, resharpen the end and try to push the broken bit out from the opposite side. When you are happy that all the pivot holes are clean, return the parts to the tray.

If the pivot holes are straight into the metal (i.e. not jewelled), check that they are still round by examining them with a magnifying glass. If they are not, it may be necessary to close up the hole with a staking tool or, if they are very worn, rebush them.

To close the hole, place the plate on the staking tool with the inside surface facing up and, with a round-ended stake, lightly tap around the pivot hole (being careful not to hit the hole itself as this may make it larger) until it is a little smaller than required for the pivot. Then using a suitably sized burnisher, open the hole so that the pivot just fits without friction. This is something that can only really be judged by trial and error, so take it slowly. If the hole is too large to be closed up by staking, a new bush may need to be inserted. This consists of a small brass disk with a hole in the middle. The old pivot must be drilled out. Then, using a reamer working from the inside, open the hole until the bush is just too big for the hole. The reamer should have produced a slightly conical hole in the plate. The bush can then be positioned over the hole (from the inside) and, using a flat-faced stake, tapped into place. Once you are happy that it is correctly positioned, open up the hole with a burnisher, as above.

Cleaning the wheels and pinions

Examine each wheel to make sure that it is flat and that the pivots are straight. If they are, clean both pivots by holding a piece of peg wood so that the pivot sinks into the blunt end and rotate it a few times. If the pinion vanes have dirt between them, this needs to be cleaned out. Sharpen a piece of peg wood to a flat point and push it along each of the vanes until all the dirt has been removed. A rub with an old (clean) tooth brush can remove most general dirt. Do not use too much force as this can distort the wheel.

Reassembling the watch

Assuming that there are no parts missing and any necessary repairs have been made, it is now time to reassemble the watch. First put the wheels in place; if each is held by a separate plate, fit them one at a time. You may need to check in which order they are put back. Start with the one whose wheel is nearest the main plate (this is when you might need to refer to any photographs you took at the start). If you are still not sure, check in a book on watches. If the wheels are all held by one plate, then you will need to fit all of them before the plate can be put in place. Hold this lightly in position (do not try to screw down yet), and check that all the pivots pass through the appropriate pivot holes (top and bottom). You may need to use a pair of tweezers to move the wheels until the pivots have engaged with their pivot holes. When you are happy that all the pivots are in place, apply a little pressure to the centre wheel. If everything is correctly in place, they should all rotate freely. Then, and only then, screw in each screw; do not tighten them fully at this stage but stop as soon as you can feel some resistance. Make sure that all the wheels can still turn freely. Once you are happy that everything is correctly positioned, tighten the screws.

Now fit the spring barrel in place. Position the barrel and plate at an angle so that the barrel goes underneath the centre wheel. Bring the spring barrel bridge down onto the main plate and make sure that it is flush. When you are happy that it is in place, fit the screws and screw them home. Never overtighten them, and always use the correct sized blade otherwise you may damage the head. Light pressure applied to the spring barrel should cause the complete train to move freely. If not, you may need to loosen the screws and check that the barrel is correctly positioned.

Now fit the escape wheel, carefully placing this the correct way up into its bottom pivot. Then place the bar over it so that the top pivot engages with the pivot hole. Insert the screw and lightly tighten. Make sure that both the bottom and top pivots are correctly engaged (check with a magnifying glass), then when (and only when) you are happy that the escape wheel can rotate freely, tighten the screws.

Never wind the watch in this state (with no balance fitted), as this will cause the escape wheel to rotate very fast, which in turn can damage or break the pivots.

Before fitting the balance, lightly oil all the pivots, including the bottom balance pivot. Do not over-oil as this can lead to a build-up of dust. Use an oil pin (a tool specially designed for the job) and transfer a drop to each pivot; aim to only put enough to partly fill the indentation around the pivot. Because a cylinder movement works by allowing the teeth of the escape wheel to be in constant contact with the cylinder, a small amount of oil should be applied to these.

Reassemble the balance and balance cock. Fit the hair spring back through the hole and position it at the same distance as before. Refit the pin and carefully press it home with a pair of tweezers. Make sure that it is correctly positioned between the regulator pin and keeper; turn this through 90 degrees so that it is prevented from escaping.

Now carefully lift the balance and balance cock, and turn it over so that the balance is dangling down (be careful not to let the balance catch on anything or be shaken too much, as this could distort the spring). Place the bottom pivot into the lower pivot jewel, which usually means that the balance needs to be put partly under the centre wheel. Then line the balance cock up so that the mounting lugs on the bottom drop into the holes in the main plate. Keeping the balance level with the main plate, very carefully push the balance cock down until it is flush with the main plate. Make sure that as the balance cock is pushed down, the top pivot engages with the jewel in the balance cock. If the balance is correctly positioned, the balance should swing freely when lightly shaken. When you are happy that it is correctly positioned, you can screw the balance cock in place. As you tighten the screw, it is a good idea to give the balance a light shake so that it is oscillating to make sure that everything is in the correct position as the screw is tightened.

If required, refit the Geneva stop work. It is important to understand how this works. You will find that before you dismantled the stop work, it was resting with the spur stopped against the 'blind' arm of the cross (assuming that the spring was in good order), with some tension in the spring. You need to wind the spring a turn or so with the key before you fit the stop piece. Now fit the stop piece in the same position it was before you removed it, resting against the blind arm of the cross. Place the winding square on a hard surface and, using a stake with a large enough centre hole, tap the stop piece over the barrel arbor. If it is correctly fitted, you should be able to wind the watch until the stop work prevents any further winding.

At this point, if all has gone well, you should have a working watch. Using a key, wind the watch up (at this stage only turn it a couple of times). Then shake the watch and it should start ticking. If it does, then proceed to wind it fully. Always be cautious when winding up a watch, especially one that has just been cleaned. You should feel the spring become stiffer as it approaches the end (unless stop work is fitted); it is always better to stop before it is fully wound than to overwind it, which may damage the spring or another part of the watch. Turn the movement over and make sure that it still ticks. Try it in several positions. Place it somewhere safe and check it after twenty-four hours; if it is still running, then you can fit the dial.

After a little while, you will develop a sense as to whether a particular watch will run, or will stop soon. You will know from the way it ticks and how the balance swings.

Fitting the dial

Push the setting pin through the centre wheel and place the movement face up with the setting square resting on a hard surface. Place the cannon pinion on it and push it fully home; if the motion work is in place, be careful that the teeth of the cannon pinion engage with the motion work before you push it home. Fit the motion work and dial washer. Fit the dial and, depending on the fixing method, either tighten the side screws or rotate the cut-away screws so that they engage with the dial feet.

Fitting the hands

If the hands are of blue steel, they only need to be carefully wiped with a clean piece of kitchen towel and fitted. If they are gold, they may need to be cleaned with metal polish. Lay the hand on a flat surface and wipe along the length with a piece of kitchen towel soaked in polish. Do not wipe towards the centre mounting, as it is very easy to catch the point and bend the hand. Finish by cleaning it with a fresh piece of kitchen towel.

Once cleaned, fit the second hand (assuming there is one); press it down with the back of a pair of tweezers until it is almost flush with the dial, making sure it is not in contact at any point. If it is, this may cause the watch to stop and can also leave marks on the dial. Next fit the hour hand, pressing it down with a pair of tweezers. Move the hour hand, using the winder (never try to push hands as this will cause them to bend or break), or a key on the setting square, so it is pointing to 12 o'clock, then fit the minute hand so that it is in line with the hour hand and press it into place with tweezers. Wind the hands round a full twelve hours to check that they do not catch on each other or on the second hand (if fitted). Always move them in a clockwise direction to prevent possible damage to the movement.

Regulation

Once the watch is working, you will need to bring it to 'regulation'. This involves adjusting the regulator arm on the balance cock backwards and forwards until the watch keeps good time. The regulator arm usually moves over a scale engraved on the balance cock, usually marked F (fast) and S (slow) but sometimes R and A (retard and advance). As the arm is moved this in turn moves the curb pins along the hair spring, so increasing or decreasing its effective length. The effect of changing the length of the hair spring is to change the rate of rotation of the balance (the beat); the longer the hair spring, the slower the beat, the shorter the spring the faster the beat.

Things that might stop the watch from working

Like all mechanical devices, watches do not always work as intended. This may be due to a number of reasons. There may be a damaged or badly fitted part, or too much friction in the train, or any number of other faults. One problem is a lack of power to the escape wheel. Let down the main spring as before and remove the balance assembly. With a key, wind the watch a couple of 'clicks' (any more could damage the escape wheel pivots). The escape wheel should rotate freely. If it does not, check that there are no bits trapped in the leaves of the pinions or between the teeth on the wheels. Another cause of a sluggish train may be distortion to one of the fixing bars. Check each bar by slackening off the screw that holds the bar in place (tighten the previous screw before checking the next one), and if the train becomes free, then this indicates that the bar might be bent. Dismantle and check the bar against a flat edge. You can straighten a bar by applying a small amount of pressure in the opposite direction. Reaseemble and check that the train is running freely when the screw is tight.

THE ENGLISH FUSEE FULL-PLATE LEVER

The lever movement was invented in the 1750s by Thomas Mudge, and developed into English fusee table roller lever during the first quarter of the nineteenth century. This style of movement came to dominate English watch making until the end of the nineteenth century. They are usually found in silver or gold double-bottom cases. The rear cover springs open when the pendant button is pressed, to reveal the winding square. Hands are usually set from the front by turning a key on the end of the cannon pinion.

There are many variations on the lever (Savage two-pin, Massey etc.), but most of them differ only in the detail of the impulse mechanism.

I describe here a typical English-made full-plate table roller lever pocket watch of the mid- to late nineteenth century (see figure 35).

Figure 35 Silver-cased English fusee lever

Removing the movement from the case

This type of watch comes in three main case styles: open-faced, half-hunter and full hunter. With the half- and full hunters, the front is usually opened by pressing the pendant button. Inside there is normally an inner bezel and a crystal. Open the inner bezel with a case knife then, using a pin pusher or pieces of steel wire just slightly smaller in diameter than the bezel pin, push the bezel pin through. In some cases, you may need to tap the end of the wire with a hammer in order to push the pin out. Once this has been extracted, the bezel and movement can be removed from the case.

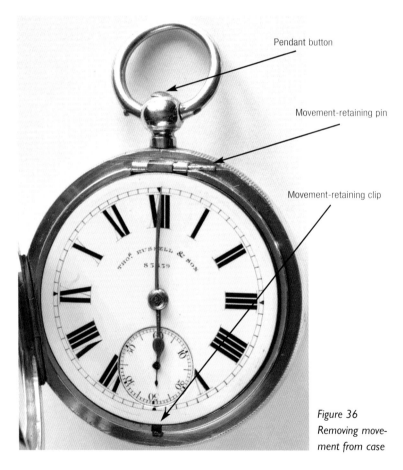

Pendant button

Movement-retaining pin

Movement-retaining clip

*Figure 36
Removing move-
ment from case*

In open-faced watches, the movement is held in place by a hinge at 12 o'clock (separate from the bezel hinge) and a spring clip at 6 o'clock. Therefore to remove the movement, open the bezel with a case knife, and then push out the pin at 12 o'clock; the movement can then be removed completely from the case.

Having successfully separated the movement from the case, you can now remove the hour and minute hands. This is best done using a purpose-made hand remover, designed to lever the hands up without damaging the dial. Press the button on top of the tool, which will cause the jaws to open. Press it down over the hands (the centre section is spring loaded) and release the button. The jaws should close under the hands. Then press the whole handle down (not the button); this should cause the jaws to lever up the hands. Position the remover so that it does not put any strain on them. The second hand can be carefully removed by holding a craft knife near the stem and levering it up; hold your finger over it too so that it does not fly off.

Removing the dust cover

Figure 37
Removing dust cap

Most watches of this type have a dust cover which encases the rear of the movement. As its name implies, it is designed to protect it from dust and damage (see figure 37). To remove this, push the curved bar clockwise using the small nib projecting from the middle. It should slide round and release the cover. This can then be lifted clear. In some watches you will find a rim dust cover. This only fits round the edge of the movement (between the plates) and is held in place by two half-headed screws. To remove this, turn the two screws until they are clear of the ring and then lift it clear.

Removing the balance

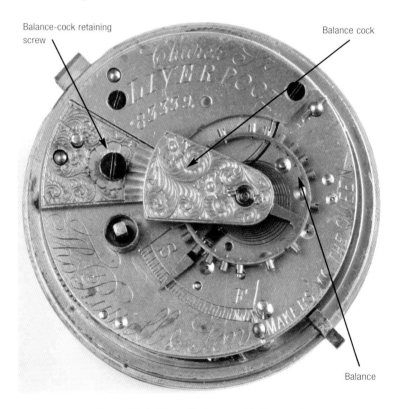

Figure 38 Position of balance and balance cock

The balance is held in place by the balance cock, which in turn is fixed in place by a single screw. Hold the movement face down and level, then undo the screw and remove it. Using a strong pair of tweezers carefully lift the balance cock clear of the balance, always trying to lift it straight up so as to avoid breaking the balance pivots. Be careful to disengage the curb pins from the hair spring if the regulator is a London quadrant type (the regulator is an arm extending from the balance cock).

The hair spring can be mounted either above or below the balance. If it is above, it will be pinned to a separate steel support, which in turn is fixed to the bottom plate with a screw. Remove the screw, and then lift the balance assembly away from the bottom plate.

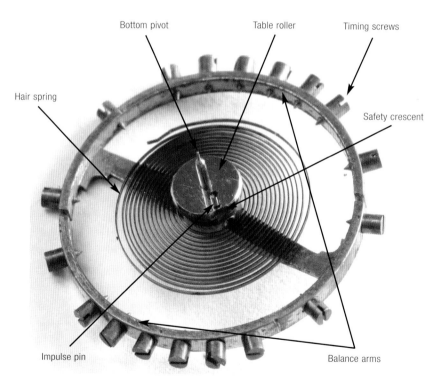

Figure 39 Balance assembly

If the hair spring is mounted below the balance it will be pinned to the hair spring stud by a brass pin. Remove the pin (be careful as it is easy to slip and bend the hair spring) and feed the spring through the stud until it is free. Lift the complete balance assembly away from the bottom plate, making sure that the hair spring is free of the curb pins on the regulator arm.

Removing the dial

The most common method of retaining dials on this style of watch is with tapered brass pins (usually three), which pass through holes in the ends of the dial feet. The dial feet should be spaced fairly evenly round the dial. To remove the pins either pull them out with a strong pair of tweezers or, if this proves to be difficult due to the tight fit of the pin, get hold of the end with a pair of cutters (only apply enough pressure to grip it, not to cut through) then pull back on the handles and this should lever the pin out. Be aware that sometimes the dial feet may have become bent. If this is the case, you may need to lift the dial on one side

Dial-retaining pin

Figure 40 Dial-retaining pin

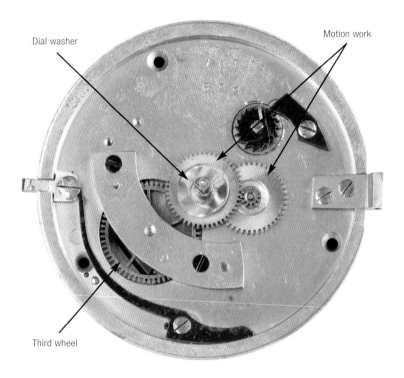

Figure 41 Under-dial view of motion works

first before you can remove it completely. Be very careful about bending the dial feet, as this may cause damage to the dial face; it is much better to leave them as they are.

In some watches (usually older ones), the dial is first fixed to a dial plate using pins, and this in turn is fixed to the top plate. Removal is exactly the same as above.

With a pair of tweezers, remove the motion work and, if present, the dial washer.

Cleaning the dial

If the dial is undamaged, it only needs to be cleaned in soap and warm water (not too hot, otherwise the enamel may be damaged). After cleaning, dry it off with a paper towel. If the dial has hairline cracks, these can sometimes be made to look less obvious by soaking in a mild bathroom cleaner (Bath Power by OzKleen seems to work well). It is normal for the numbers to be an integral part of the enamel dial and they are therefore unlikely to be removed by any cleaning. However if there is other writing on the dial (maker's name etc.) care must be taken to ensure that it is not removed during cleaning, so check a small area first. Once cleaned, put the dial safely to one side.

Removing the cannon pinion

Cannon pinion

Figure 42 Cannon pinion

After the motion work has been removed, the cannon pinion is next. This is held in place by a friction fit. To remove it, you need to take hold of it with a pair of small pliers and, with a twisting action, pull it firmly away from the front of the watch. Be careful not to apply too much force as you may distort it. If it proves to be very stiff, a small amount of oil should be applied and left overnight to work its way in.

Letting down the main spring

Before removing any further parts, it is *very* important that you release the tension in the main spring. If you find that there is none this may be due to a broken spring and/or a broken or lose fusee chain. If this is the case, you can skip the next section.

Hold the movement face up and you should see the third wheel bridge held in place by two screws. While keeping pressure on the bridge, undo and remove both screws; it is very important that

Figure 43 Position of set-up ratchet and paw

77

you do not allow the bridge to become detached from the plate. Place the correct sized key over the winding square and hold it firmly with your thumb and forefinger. Then carefully lever off the bridge and remove the third wheel, while keeping a firm hold of the key (remember there is quite a lot of energy stored in a fully wound spring). You should find that the key now wants to turn; allow it to do so, all the while keeping it firmly between your fingers until the spring has completely run down.

Slacken off the screw that holds the set-up paw against the set-up ratchet and push the paw away, which should release the last bit of tension in the spring. You can now pull the fusee chain out so that it unwinds from the spring barrel. Do not pull too hard as it may be caught somewhere in the movement. If so, wait until the

Spring-barrel bridge-retaining screws

Figure 44 Position of spring-barrel bridge-retaining screws

spring barrel has been removed before removing it. If, however, it unwinds easily, when it is free carefully unhook it from the spring barrel, and then do the same at the other end from the fusee. Now remove the set-up ratchet wheel by pulling it clear.

Turn the movement over and remove the two screws that hold the spring-barrel bridge in place, and then lift this clear. Now lift the spring barrel away from the movement.

Dismantling the plates

Most English fusee levers are held together by brass pins, although a few (usually from the very late nineteenth century) use screws.

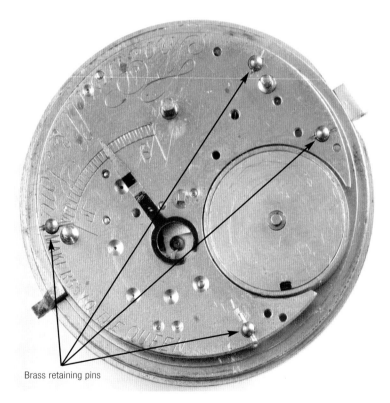

Brass retaining pins

Figure 45 Position of bottom-plate retaining pins

To remove the bottom-plate retaining pins, use a pair of pliers to push them through by placing one side of the jaws against the thinner end of the pin and the other jaw against the other side of the pillar above the pin (the jaws will therefore be at a slight angle). Press the jaws together and the pin should pop out. The ones under the balance cock or spring barrel will be only just longer than the width of the pillar head. To remove these, try the same method as for the longer ones above. If this fails, try removing them with a piece of steel wire that has been filed sharp at one end and hardened. By applying pressure to the thinner end of the pin (this is usually the end pointing towards the centre of the movement), it should be possible to push the pin out.

Once all the pins have been removed, you can remove the bottom plate. You will need to lift the plate clear of the four pillars carefully. Be very careful that you do not catch the lever between the top and bottom plates. Because one end of the lever engages with the balance, it protrudes over the bottom balance pivot, which means that when the plates are separated it must remain with the bottom plate or you might snap the lever top pivot. The other part to keep look out for is the ratchet for the maintaining power. This is held against the ratchet wheel (usually steel) on the fusee. It is positioned between the two plates and held there by a pivot top and bottom. Because it is under tension from a spring, it can fly off during disassembly if you are not careful.

Having separated the plates, remove the centre, second and escape wheels, then remove the fusee.

Cleaning the plates

Before cleaning the plates, it might be worth removing the fusee stop work spring, as this can easily be bent or damaged by cleaning. This is held in place by a single screw; undo this and lift the spring away from under the fusee stop. Also remove the fusee ratchet-paw spring on the top plate.

The best way I have found to clean the plates and other parts

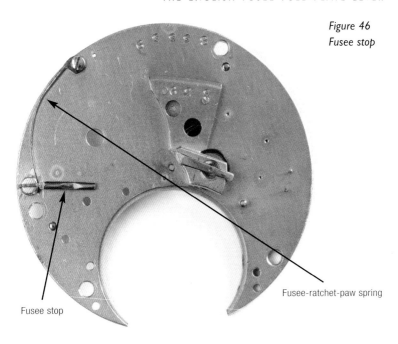

Figure 46
Fusee stop

Fusee-ratchet-paw spring

Fusee stop

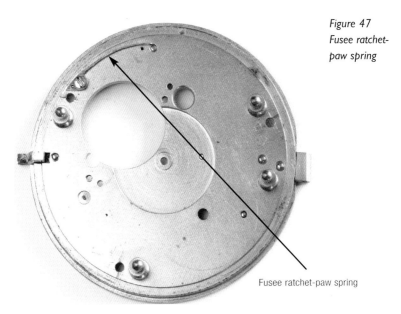

Figure 47
Fusee ratchet-
paw spring

Fusee ratchet-paw spring

is to use a metal polish. Some older books on watch repairing advocate the use of benzine, but this is not only difficult to obtain but is also a poison and should not be used. Using kitchen towel and metal polish, rub the surface until all the dirt, marks and grease have been removed. Be careful not to damage or move the lever banking pins. After polishing the plates with a clean towel, the next step is to clean the pivot holes. This is done using a length of peg wood (see list of suppliers in appendix A), whose end has been sharpened to a point with a craft knife. Insert this into the pivot hole and rotate a few times, repeat until the wood comes out clean. Do not be tempted to reuse the peg wood on another pivot hole without resharpening it, as this can lead to the point breaking off and sticking in the hole. If this happens, resharpen the end and try to push the broken bit out from the opposite side. When you are happy that all the pivot holes are clean, return the parts to the tray.

Figure 48 Third wheel bridge

Do not forget to clean the third wheel bridge as well.

If the pivot holes are straight into the metal (i.e. not jewelled), check that they are still round by examining them with a magnifying glass. If they are not, it may be necessary to close up the hole with a staking tool or, if very worn, rebush them (see page 157).

Cleaning the wheels and pinions

Examine each wheel to make sure it is flat and that the pivots are straight. If they are, clean both pivots by holding a piece of peg wood so that the pivot sinks into the blunt end and rotate it a few times. If the pinion vanes have dirt between them, this needs to be cleaned out. Sharpen a piece of peg wood to a flat point and push along each of the vanes until all the dirt has been removed. A rub with an old (clean) toothbrush can remove most general dirt. Do not use too much force as this can distort the wheel.

Checking the spring

Figure 49 Spring barrel with cover removed

One of the more common problems with pocket watches is a broken spring. To check for this, remove the spring barrel cover by inserting a screwdriver into the slot and carefully lever it up.

Check that the unbroken spring is still connected at both ends. Either the outer end should have a hole which hooks over a pin on the inside of the spring barrel or it should have an oblong piece of steel riveted to it which fits a similar shaped hole in the barrel. The inner end should fix onto a hook on the barrel arbor. If this all looks to be all right, then you need to apply a small amount of oil and replace the cover. This fits into a small groove. It may need to be pressed back into place with a pair of brass-faced pliers and should be flush with the barrel when properly fitted.

Checking the fusee

Even if the fusee appears to be working correctly, it is good practice to dismantle and clean it. Assuming it has maintaining

Figure 50 Fusee

Figure 51 Fusee (underside)

power, the fusee consists of three main parts. The main body consists of the brass fusee cone attached to the steel winding square (see figure 50). Below this is the steel ratchet wheel (see figure 53), which engages with the ratchet on the underside of the main body. Finally, the brass main wheel incorporates the maintaining-power spring (see figure 54). All three parts are held together by a pin which passes through the lower part of the shaft from the main body. To dismantle it, push out the pin and then remove the collar. Next remove the main wheel and finally the ratchet wheel. Sometimes the pin can be difficult to remove; if this is the case you may need to use a short piece of hardened steel with a sharp point. Hold the pointed end against the pin and tap it lightly until the pin is forced out. Once enough of the pin is clear, use cutters to remove it completely.

*Figure 52
Fusee (under-
side of fusee
cone showing
internal
ratchet)*

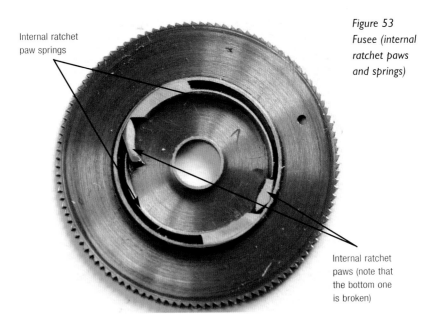

Internal ratchet
paw springs

*Figure 53
Fusee (internal
ratchet paws
and springs)*

Internal ratchet
paws (note that
the bottom one
is broken)

Figure 54
Fusee (main-
taining spring)

First examine the ratchet wheel on the underside of the fusee cone for wear. If it looks all right, proceed to the ratchet, which consists of two paws (on some older types, there may be only one) that engage with the ratchet wheel and are held against it by springs, which are an integral part of the wheel. The paws are riveted into this and are free to move. If the watch has been badly treated, these may be damaged or bent by the fusee being forced backwards. Replacement paws can be made but, because of their size, this is quite difficult. You would be better to try and replace a damaged one with one from a scrap fusee. The same is true if the ratchet wheel is badly damaged. The maintaining power spring is much less likely to be faulty, but sometimes the small pin that connects to the ratchet wheel, or the hole in the ratchet wheel that engages with it, can become worn. If this happens, the pin can be replaced by tapping the old one out and fixing a new one in place. This will need to be filed to the correct height. If the hole on the ratchet wheel is worn, it can sometimes be cleaned up by opening it very slightly with a reamer.

If the fusee parts are all in order, then just give the inside a clean and oil and reassemble the parts. Finally reinsert the pin to hold it together.

Fusee chain

Figure 55 Fusee chain

The fusee chain consists of dozens of links hand-riveted together to form a very strong and flexible chain with a hook at either end. It is designed to connect the fusee to the spring barrel. When the watch is wound up, the chain is transferred from the spring barrel onto the fusee and, as the watch runs down, it winds back onto the spring barrel. If it is fine and both the hooks are good, then it only needs a light oil. If it is stiff and discoloured, it needs a little more attention. Take a length of peg wood about 5mm in diameter and cut a slot in one end. Hold it upright with the chain held across the slot. Place a few drops of oil on the chain, then pull it backwards and forwards through the slot, moving along its entire length. Do this until the chain feels reasonably free. Wipe off the excess oil.

If the chain has broken (a very common cause of a fusee lever watch not running), then it needs to be either replaced or repaired (see page 163). When choosing a replacement, it is important that it is of similar thickness and length. If the replacement chain is thicker than the original one, it may not fit the groove cut in the fusee; if it is too thin, there is the possibility that it will not be strong enough for the spring and may subsequently break. If the original chain is missing, then you can check whether a replacement one is the correct length by fixing the hook at the start of the fusee and winding it onto it until you reach the top (just below the 'stop'). There should be a minimum of 3cm of the chain left to attach to the spring barrel. A bit more is not a problem, but any less will probably lead to the chain breaking when the watch is first wound, as the stop will not engage.

Before you reassemble the watch, make sure you fix the stop work spring in place. The flattened end should go under the stop work and, when screwed back in place, it should lift it clear of the bottom plate. If you press down and let go, the stop work should spring up again. It is designed to be pushed flush with the bottom plate when the fusee chain reaches the last turn on the fusee. The end of the stop work should then engage with the stop on the fusee and prevent it from turning any more.

Reassembling the watch

Assuming that there are no parts missing and any necessary repairs have been made, it is now time to reassemble the watch. Take the top plate and place the fusee paw in its pivot hole but turned through 180 degrees so that it points in the opposite direction (held by spring). Put the plate aside.

Now take the bottom plate and, holding it upside down, place the lever in its position (apply a small amount of oil to the faces of the pallets). Then place the escape wheel and second wheel in their respective locations. Put the fusee in place and then finally put in the centre wheel (see fig 57). Then turn the top plate over and bring it carefully down over the bottom plate, allowing the centre wheel pin to pass through the appropriate hole. Make

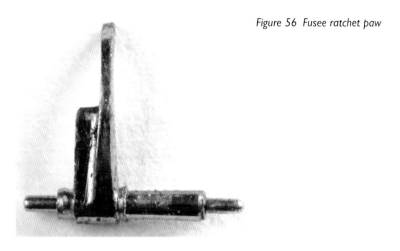

Figure 56 Fusee ratchet paw

Figure 57 Bottom-plate assembly

sure the two plates are aligned (the pillars lined up with their corresponding holes). Slowly put the two plates together so the pillar heads start to go through their corresponding holes. At this point check that everything is still in place, and push them together until the lever pivot and fusee paw are touching the top plate. Using a pair of tweezers, manoeuvre these two parts until the pivots drop into their appropriate pivot holes. Now push the bottom plate down until it is in the correct position. Check carefully that no parts are trapped and everything is in the correct position. Pin the plate in place with the four brass pins (remember that the two to the left of the spring barrel should be short).

Turn the movement over and fit the third wheel bridge, but not the third wheel because the bottom pivot for the fusee is on this plate and it also protects the second wheel pivot (where the second hand fits).

Fitting the spring barrel

Place the spring barrel with the set-up arbor through the top plate. Fit the spring barrel bridge and fix it in place with the two screws.

Fitting the fusee chain

Flip the fusee ratchet paw over so that it engages with the ratchet wheel. Rotate the fusee both ways with a key and check that it moves correctly. Leave the fusee so that the attachment point for the fusee chain is on the outside edge. Take the fusee chain and, holding the movement edge on, drop the chain past the spring barrel behind the two pillars so that the end comes out by the fusee (note that a fusee chain has a different type of hook at either end; the pointed one fits on the spring barrel). Using tweezers, hook the end of the fusee chain onto the fusee; there should be a slot cut into the fusee, and in the middle of this a steel pin to which the hook must be attached. Pull the chain until the fusee stops rotating.

Using a key attached to the set-up arbor, rotate the spring barrel anticlockwise until the small slot for the fusee chain is on the outside. Hook the free end of the fusee chain into the slot. Keeping the chain away from the movement (with your finger), rotate the spring barrel with the key and feed the chain onto it (try to keep the chain parallel and away from the bottom of the spring barrel). When all the chain has been wound onto the barrel, keep pressure on the spring barrel, remove the key, fit the set-up ratchet over the arbor and engage the paw. Then, push the ratchet round a few notches with a screwdriver, keeping the paw engaged. Tighten the paw screw to hold it in place (be very careful not to slip). The fusee chain should be under tension.

Carefully remove the third wheel bridge and fit the third wheel, refit the plate and screw it in place. Make sure that before you fit the screws, all the pivots (especially the escape wheel pivot, if it is located on the bridge) are correctly located in their pivot holes. Then tighten the bridge screws.

Fitting the balance

Before the balance is fitted, oil the pivots, including the top and bottom balance pivots; do not over-oil them as this can lead to a build up of dirt. Use an oil pin (a tool specially designed for the job) and transfer a small drop to each pivot; aim to only put enough to partly fill the indentation around the pivot.

Then place the balance pivot on the bottom pivot hole and, depending on whether the hair spring is mounted above or below the balance, proceed in one of two ways. If it is below, feed the end of the hair spring through the hole on the hair spring stud with a pair of tweezers. Make sure that the outer turn of the hair spring passes through the index curb pins. Feed the hair spring through until the impulse pin is roughly in line with the centre of the lever. Fix it in place with a brass pin, and then fit the balance cock.

If the hair spring is above the balance, position the spring support so that the pin fits into the hole and fix it with the screw. Then place the balance cock in place, making sure that the index lever curb pins engage correctly with the outer turn of the hair spring.

With both arrangements, check that the balance swings freely and also that, when it comes to a stop, the lever is in a central position – i.e. it is midway between the banking pins. If it is not you will need to adjust the hair spring by loosening the brass pin and moving the spring backwards or forwards until the lever is central. This is straightforward if the banking pins are at the rear of the lever. If they are towards the front, it is more difficult to judge the correct position and you may need to listen to the movement when it is running to determine the correct position.

Once everything is set up, you can screw the balance cock in place. It is a good idea to give the movement a gentle shake so that the balance is swinging, then slowly tighten the balance cock screw until it is fully home. Check that the balance still swings freely, trying it in several positions.

Using a loupe, examine the hair spring, which should be level with the bottom plate and clear of both the balance cock and the bottom plate. It should also be symmetrical around the balance pivot. When the balance is swung, the spring should expand and contract evenly. If it does not, use a pair of tweezers to adjust it into the correct position.

Note on screws

In high-quality watches you might come across screws with one, two, three or more small dots punched into the heads. They are matched by dots punched into the movement next to the screw holes. When you reassemble the watch, make sure that the dots are correctly matched.

Winding up

At this point you are ready to wind the watch. A bit of care is required as the fusee chain will not necessarily be aligned with the grooves on the fusee. Fit a key to the winding arbor and slowly wind it in an anticlockwise direction. Keep an eye on the chain and make sure that it is feeding into the grooves correctly. If it is not, use tweezers to move it up or down the spring barrel to realign it. If during winding the chain slips off the fusee, stop winding, unwind the fusee and start again. As the chain comes to the top of the fusee, it should push the fusee stop work up against the bottom plate; this then meets the stop piece on the fusee and should prevent any further rotation. Keep a very close eye on the stop work as you reach the end to make sure that it functions correctly. If you reach the end of the chain before the stop work has engaged, do not wind any further (stop before the chain is pulled off the spring barrel, as this could damage it). If it does not engage, you will need to dismantle the movement and adjust the stop work.

If all goes well, a shake of the watch should set it going. Put it aside in a safe place and leave it for twenty-four hours; if it is still running, then you can fit the dial.

After a little while, you will develop a sense of whether a particular watch will run well or stop soon. You will know from the way it ticks and the way the balance swings.

Fitting the cannon pinion

The cannon pinion is held in place by friction. Place it over the centre wheel pinion and press it down with pliers. You should be able to rotate it forwards and backwards but there should be some friction or it will not rotate when the centre wheel moves.

Fitting the dial

Fit the motion work and dial washer. If the dial is fitted to a dial plate, fit this to the dial and fix it in place with brass pins. Then fit the dial and plate to the movement. If the dial is not fitted to a dial plate then fit the dial directly on the movement. In both cases, fix it with brass pins.

Fitting the hands

If the hands are of blue steel, they only need to be carefully wiped with a clean piece of kitchen towel and fitted. If they are gold, they may need to be cleaned with metal polish. Lay the hand on a flat surface and wipe along the length with a piece of kitchen towel soaked in polish. Do not wipe towards the centre mounting as it is very easy to catch the point and bend the hand. Finish by cleaning with a fresh piece of kitchen towel.

Once cleaned, fit the second hand (assuming there is one). Press it down with the back of a pair of tweezers until it is almost flush with the dial, making sure it is not in contact at any point. If it is, this may cause the watch to stop and can also leave marks on the dial. Adjust the minute hand arbor so that the square allows the hand to point at 12 o'clock; next fit the hour hand (aligned with 12 o'clock), and press it down with a pair of tweezers. Now fit the minute hand and press it down with the correct sized key. Using a key, rotate the hands through twelve hours to make sure they do not catch (always move them in a clockwise direction to prevent possible damage to the movement).

Regulation

Once the watch is working, you will need to bring it to 'regulation'. This involves adjusting the regulator arm backwards and forwards until the watch keeps good time. The regulator arm moves over a scale engraved on the bottom plate, usually marked F (fast) and S (slow). As the arm is moved this in turn moves the curb pins along the hair spring, so increasing or decreasing its effective length. The effect of changing the length of the hair spring is to change the rate of rotation of the balance (the beat); the longer the hair spring the slower the beat, the shorter the spring the faster the beat.

Things that might stop the watch from working

Like all mechanical devices, watches do not always work as intended. This may be due to a number of reasons. There may be a damaged or badly fitted part, or too much friction in the train, or any number of other faults.

Do not assume that all the parts found in a watch movement started life with that watch. It is quite possible that someone in the past has replaced a damaged or missing part in order to get the watch to work, which may not have been successful. The replacement part may therefore be the cause of the movement not working.

If the balance does not swing freely, this can be caused by a number of problems.

- Check that the hair spring is level and not catching on the bottom plate or balance cock.
- Check that the pivot jewels are clean and not damaged.
- If the hair spring is below the balance, is the balance catching on the hair spring stud? If it is, you can carefully reduce the height with an Arkansas stone.
- Burnish the top and bottom pivot with a Jacot tool.
- If the safety pin on the lever is bent forward and rubbing on the table roller, dismantle the movement and remove the lever, then carefully bend the pin backwards. Reassemble and test it.

If when the balance is shaken, the lever does not remain engaged with the safety roller, it is usually because the guard pin, which passes through the crescent on the roller, is bent so that it can 'jump' out of sync with the balance. To fix this, dismantle the movement and remove the lever, then carefully bend the safety pin forwards a small amount. Reassemble and test it.

Another problem may be a lack of power to the lever. Remove the balance and balance cock then, using a piece of sharpened peg wood, move the lever from side to side. If it is moved a small amount, it should snap back when you let go. If you move it a bit more, it should snap over to the other side. If this does not happen, the problem lies in the train. Let down the tension in the spring and disassemble the movement. Reassemble it without the lever, then wind it up a little, so that the escape wheel can rotate freely. Do not wind it up too much, as the wheel pinions could be damaged if they are allowed to run too fast. If it does not turn freely, check that there are no bits trapped in the leaves of the pinions or between the teeth of any of the wheels.

Another cause of a sluggish train may be a distorted fixing plate. Check that both the top and bottom plates are flat; if not, they need to be flattened. You can straighten them by applying a small amount of pressure in the opposite direction. Reassemble and check if the train is free-running.

THE ENGLISH THREE-QUARTER-PLATE CENTRE SECONDS CHRONOGRAPH

This was a popular style that appeared in large numbers towards the end of the nineteenth century. It is based on a version of the English three-quarter-plate fusee lever.

Figure 58 English three-quarter-plate centre seconds chronograph

Removing the movement from the case

This type of watch really only comes in one style of case: open-faced with an inner and outer back cover. The front is usually hinged at 9 o'clock. Using a case knife at 2 o'clock lever the front open. The movement is held in place by a hinge at 12 o'clock and a pin at 6 o'clock. Use a pin pusher or piece of steel wire just slightly smaller in diameter than the hinge pin to push this through. In some cases, you may need to tap the end of the wire with a hammer, in order to push the pin out. Once this has been removed, the movement can be removed from the case by carefully pushing it from the rear near the top.

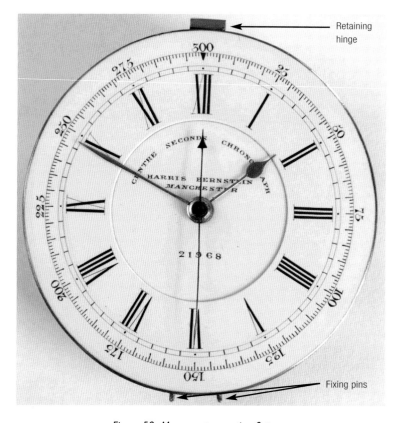

Figure 59 Movement mounting fixtures

Having successfully removed the movement, you can now remove the centre second, hour and minute hands using a purpose-made hand remover. This is designed to lever the hands up without damaging the dial. Press the button on top of the tool; this will cause the jaws to open. Press it down over the hands (the centre section is spring loaded) and release the button. The jaws should close under the rim of the hands. Press the whole handle down, which should cause the jaws to lever up the hands and free them from the cannon pinion. Position the hand remover so that it does not put any strain on the hands.

Removing the balance assembly

Balance-cock
retaining screw

Figure 60 Balance cock

The balance is held in place by the balance cock, which in turn is fixed in place by a single screw. Hold the movement face down and level, then undo the screw and remove it. Using a strong pair of tweezers carefully lift up the balance cock and balance assembly.

Removing the balance

The hair spring is retained by a brass pin through a stud. The stud may be an integral part of the balance cock or it may be separate

Note that the balance cock has been raised by a piece of paper.

Figure 61 Balance assembly

and held in place by a small screw. If it is an integral part, then push the pin out with tweezers and feed the hair spring through the hole and curb pins. If it is separate, then remove the screw and lift the stud away from the balance cock. Then carefully lift the hair spring away from the curb pins.

Removing the dial

The most common method of retaining dials on centre second fusee lever watches is via tapered brass pins (usually three) which pass through holes in the dial feet. These will be spaced evenly round the dial. To remove them, either pull them out with a strong pair of tweezers or, if this proves to be difficult due to the tight fit, get hold of the end with a pair of cutters (only apply enough pressure to grip it and not to cut through) then pull back on the handles. This should lever the pin out. Be aware that sometimes the dial feet may have become bent. If this is the case, you may need to lift the dial on one side first before you can remove the dial completely. Be very careful about bending the dial feet, as this may cause damage to the front of the dial enamel. If they are bent, it is much better to leave them as they are.

Removing the driving wheel

With tweezers, remove the motion work and, if there is one, the dial washer.

After the motion work has been removed, the drive wheel and hand-setting arbor are next. The hand-setting square consists of a thin pin with a square block on one end. This fits from the rear and passes through the middle of the centre wheel, with the end protruding through the front. The drive wheel is fitted to this. To remove the drive wheel try using the hand remover; if this does not work, you may be able to lever it off with a screwdriver.

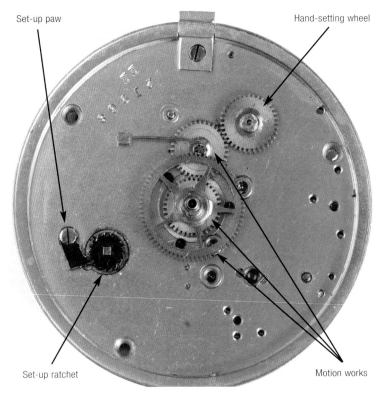

Figure 62 Under-dial view

Figure 63 Motion works

Figure 64
Hand-setting
wheel and
arbor

Cleaning the dial

If the dial is undamaged, it only needs to be cleaned in soap and warm water (not too hot, otherwise this may damage the enamel). After cleaning, dry it off with a paper towel. If the dial has hairline cracks, these can sometimes be made to look less obvious by soaking in a mild bathroom cleaner (Bath Power by OzKleen seems to work well). It is normal for the numbers to be an integral part of the enamel dial and they are therefore unlikely to be removed by any cleaning. However if there is other writing on the dial (maker's name etc.) care must be taken to make sure that this is not removed during cleaning, so check a small area first. Once cleaned, put safely to one side.

Letting down the main spring

Before removing any further parts, it is *very* important that you release the tension in the main spring. If you find that there is no tension in the spring, this may be due to a broken spring and/or a broken or loose fusee chain. If this is the case, you can skip the next section.

Fit a suitably sized key to the winding arbor and hold it firmly. Carefully remove the screw from the plate that holds the lever and

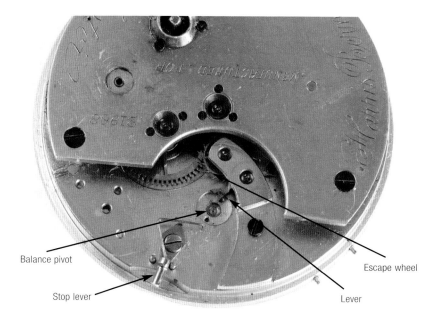

Balance pivot

Stop lever

Escape wheel

Lever

Figure 65 Movement with balance assembly removed

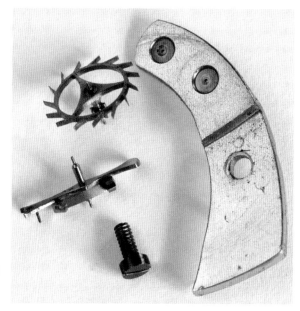

Figure 66 Lever,
escape wheel
and bridge

escape wheel. Lever up the plate while still keeping firm pressure on the key. Remove the lever and escape wheel. Allow the key to turn slowly between your fingers until the watch has run down.

Slacken off the screw that holds the set-up paw and using a piece of peg wood push this away from the set-up ratchet to release the last of the tension in the spring.

Dismantling the plates

In most English fusee lever chronographs, the bottom plate is held in place by screws, but a few use brass pins. Unscrew these (there are usually three) and remove them.

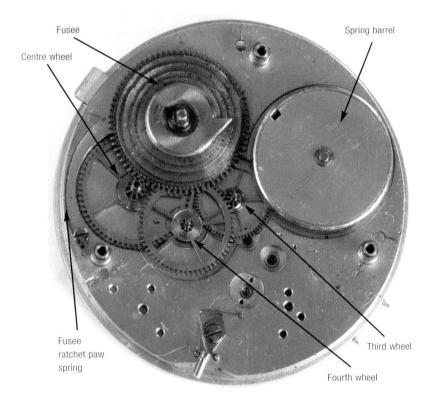

Fusee

Spring barrel

Centre wheel

Fusee
ratchet paw
spring

Third wheel

Fourth wheel

Figure 67 Top plate assembly

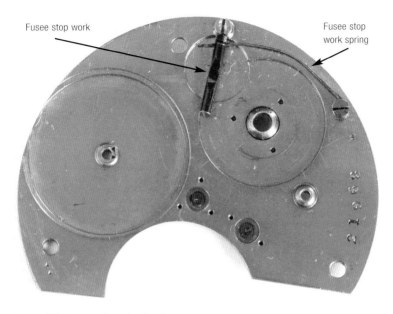

Fusee stop work

Fusee stop work spring

Figure 68 Bottom plate showing fusee stop

Figure 69 Top plate, ready for cleaning

Once all the screws have been removed, you can now remove the bottom plate. You need to keep a close eye on the fusee maintaining power paw. This is held against the ratchet wheel (usually steel) on the fusee. It is positioned between the two plates and held there by a pivot top and bottom. Because it is under tension from a spring, it can fly off during disassembly if you are not careful.

Having separated the plates, unhook the fusee chain from the fusee and unwind it from the spring barrel; remove the fusee, second, third and centre wheels and spring barrel.

Cleaning the plates

Before cleaning the plates, it might be worth removing the fusee stop work spring, as this can easily be bent or damaged. It is held in place by a single screw; undo this and lift the spring away from under the fusee stop. Also remove the fusee paw spring on the top plate.

The best way I have found to clean the plates and other parts is to use a metal polish. Some older books on watch repairing advocate the use of benzine, but this is not only difficult to obtain but is also a poison and should not be used. Using kitchen towel and metal polish, rub the surface until all the dirt, marks and grease have been removed. Be careful not to damage or move the lever banking pins. After polishing the parts with a clean towel, the next step is to clean the pivot holes. This is done using a length of peg wood (see list of suppliers in appendix A), whose end has been sharpened to a point with a craft knife. Insert this into the pivot hole and rotate it a few times. Repeat until the wood comes out clean. Do not be tempted to reuse the peg wood on another pivot hole without resharpening it, as this can lead to the point breaking off and sticking in the hole. If this happens, resharpen the end and try to push the broken bit out from the opposite side. When you are happy that all the pivot holes are clean, return the parts to the tray.

Do not forget to clean the lever and escape wheel plate as well.

If the pivot holes are straight into the metal (i.e. not jewelled), check that they are still round by examining them with a magnifying glass. If they are not, it may be necessary to close up the hole with a staking tool or, if very worn, rebush them (see page 157).

Cleaning the wheels and pinions

Figure 70
Centre, third
and fourth
wheels

Examine each wheel to make sure that it is flat and that the pivots are straight; if so, clean both pivots by holding a piece of peg wood so that the pivot sinks into the blunt end and rotate it a few times. If the pinion vanes have dirt between them, this needs to be cleaned out. Sharpen a piece of peg wood to a flat point and push it along each of the vanes until all the dirt has been removed. A rub with an old (clean) toothbrush can remove most general dirt. Do not use too much force as this can distort the wheel.

Checking the spring

One of the more common problems with pocket watches is that the spring can often break. To check this, remove the spring barrel cover by inserting a screwdriver into the slot and carefully levering up the cover. Check that the spring is unbroken and that it is still connected at both ends. Either the outer end should have a hole which hooks over a pin on the inside of the spring barrel or it should have an oblong piece of steel riveted to the end of the spring, which then fits a similar-shaped hole in the barrel. The inner end should fix onto a hook on the barrel arbor. If this all looks to be in order, then you only need to apply a small amount of oil and replace the cover. This fits into a small groove and may need to be pressed back into place with a pair of brass-faced pliers. It should be flush with the barrel when properly fitted.

Figure 71 Opening spring barrel

Figure 72 Spring barrel with cover removed

Checking the fusee

The fusee consists of a cone-shaped block of brass with a groove cut into it, along which the fusee chain winds. Even if it appears to be working correctly, it is good practice to dismantle and clean it. It is made up of three main parts. The first is the body where the fusee chain winds, which is attached to the steel winding square. Below this is the steel ratchet wheel. Last comes the brass main wheel and the maintaining power spring. All three parts are held together by a pin which passes through the lower part. To dismantle it, push out the pin, remove the collar and remove the wheel and finally the ratchet (see figures 50–54).

First examine the ratchet wheel on the underside of the fusee cone for wear. If it looks all right, proceed to the ratchet, which consists of two paws that engage with the ratchet wheel and are held against it by two springs (which are an integral part). The paws are riveted into this and are free to move. They are held against the ratchet on the underside of the fusee cone by the two springs. If the watch has been badly treated, these may be damaged or bent owing to the fusee being forced backwards. Replacement paws can be made, but because of their small size, this is quite difficult. You would be better to try and replace a damaged ratchet with one from a scrap fusee. The same is true if the ratchet wheel is badly damaged.

A common problem is that the two springs may have weakened over the years and no longer hold the paws against the ratchet with any force. To repair this, lift the spring over the paw with a craft knife and carefully push it towards the centre with a screwdriver. You want the spring to come to rest just beyond the paw. When you are happy with its position, lift it back over the paw; repeat for the other spring.

The maintaining power spring is much less likely to be faulty, but sometimes the small pin that connects it to the ratchet wheel, or the hole in the ratchet wheel that engages with it, can become worn. If this happens, the pin can be replaced by tapping the old one out and fixing a new one in place. This will need to be filed to the correct height. If the hole on the ratchet wheel is worn, it can sometimes be cleaned up by opening it ever so slightly with a reamer.

If all is well with the fusee parts then just give the inside a clean and oil then reassemble the parts. Finally reinsert the brass pin and press it home with a pair of pliers.

Fusee chain

The fusee chain consists of dozens of links, all hand-riveted together to form a very strong and flexible chain with a hook at either end. It is designed to connect the fusee with the spring barrel. When the watch is wound up, it is transferred from the spring barrel on to the fusee; as the watch runs down, the chain winds back on to the spring barrel. If the chain and both hooks are good, then it only needs light oil. If the chain is stiff and discoloured, it needs a little more attention. Take a length of peg wood about 5mm in diameter and cut a slot in one end. Hold it upright with the chain held across the slot, place a few drops of oil on the chain, then pull it backwards and forwards through the slot, moving along its entire length. Do this until the chain feels reasonably free. Wipe the excess oil from the chain.

If the chain has broken (a very common cause of the fusee lever watch not running), then it needs to be either replaced or repaired (see page 163). When choosing a replacement, it is important that this is of similar thickness and length. If the replacement chain is thicker than the original one, it may not fit the groove cut in the fusee; if it is too thin, there is the possibility that it will not be strong enough for the spring and subsequently break due to the strain. If the original chain is missing, then you can check whether a replacement one is the correct length by fixing the hook at the start of the fusee and winding it on to the fusee until you reach the top (just below the 'stop'). There should be a minimum of 3cm of the chain left to attach onto the spring barrel. A bit more is not a problem, but any less will probably lead to the chain breaking when the watch is first wound up as the stop will not engage.

Before you reassemble the watch, make sure you fix the stop work spring in place. The flattened end should go under the stop work and, when screwed back in place, should lift it clear of the bottom plate. If you press down and let go, the stop work should spring up again. It is designed to be pushed flush with the bottom plate when the fusee chain reaches the last turn on the fusee. The end of the stop work should then engage with the stop on the fusee and prevent it from turning any more.

Reassembling the watch

Assuming that there are no parts missing and any necessary repairs have been made, it is now time to reassemble the watch.

Take the top plate and place the fourth wheel in place, then fit the centre wheel, followed by the fusee and spring barrel. Then fit the third wheel. Finally refit the fusee ratchet paw spring and paw.

Check that you have refitted the fusee stop work spring, then take the bottom plate and place it in its position over the mounting pillars, first allowing the winding arbor to pass through the appropriate hole. Then make sure the two plates are aligned (the pillars lined up with their corresponding holes). Slowly put the two plates together so that the pillar heads start to go through their corresponding holes. At this point check that everything is still in place, pushing them together until the pivot of the wheels and fusee paw are touching the top plate. Using a pair of tweezers, manoeuvre the pivots until they drop into their appropriate pivot holes. Now push the bottom plate down until it is in the correct position. Check carefully that no parts are trapped and everything is in the correct position. Check that the train moves freely and then, if all is well, screw the bottom plate in place.

Fitting the fusee chain

Rotate the fusee both ways with a key and check that it moves correctly. Leave the fusee so that the attachment point for the fusee chain is on the outside edge. Take the fusee chain and, holding the movement edge on, drop the chain past the spring barrel behind the pillar so that the end comes out by the fusee (note that a fusee chain has a different type of hook at either end; the pointed one fits on the spring barrel). Using tweezers, hook the end of the fusee chain onto the fusee; there should be a slot cut into the fusee, and in the middle of this there is a steel pin, to which the hook must be attached. Pull the chain until the fusee stops rotating.

Using a key attached to the set-up arbor, rotate the spring barrel anticlockwise until the small slot for the fusee chain is on the outside. Hook the free end of the fusee chain into the slot. While keeping the chain away from the movement with your finger, rotate the spring barrel with the key and feed the chain onto it (try to keep the chain parallel and away from the bottom of the spring barrel). When all the chain has been wound onto the barrel, keep pressure on the spring barrel, remove the key and fit the set-up ratchet over the arbor and engage the paw. Then using a screwdriver, push the ratchet round a few notches, keeping the paw engaged, then tighten the paw screw to hold it in place (be very careful not to slip). The fusee chain should be under tension.

Fitting the lever and escape wheel

Using tweezers, place the lever, with a small amount of oil applied to the faces of the pallets, and escape wheel in position. Fit the plate and apply light pressure while manoeuvring the lever and escape-wheel pivots into their respective positions. Once they are correctly located, screw the plate in place.

Fitting the drive wheel

Push the hand-setting arbor (with washer) through the second wheel, turn the movement over and place the end of the square on a hard metal surface. Fit the drive wheel over the setting arbor and, using a flat-faced stake, tap it down until it is flush with the surface of the top plate. Check that when the hand-setting arbor is turned it can move; there should be some resistance due to friction.

Fitting the balance to the balance cock

Depending on the method of retaining the hair spring, either feed the end through the stud and pin it in place, or screw the stud on to the balance cock. Now fit the hair spring between the curb pins.

Fitting the balance assembly

Before the balance is fitted, oil the pivots, including the top and bottom balance pivots; do not over oil as this can lead to a build-up of dirt. Use an oil pin (a tool specially designed for the job) and transfer a drop to each pivot; aim to apply just enough to partially fill the indentation around the pivot.

Lift the balance assembly with a pair of tweezers and place it roughly in position, making sure that the bottom pivot is engaged, then lower the balance cock so that it engages with the top plate, making sure that the balance top pivot is correctly engaged. Slowly press the balance cock down while checking that the balance can swing freely. When the balance cock is fully positioned, give the movement a gentle shake so that the balance is swinging, then slowly tighten the balance cock screw until it is fully home. Check that the balance still swings freely – try it in several positions.

Check that when the balance is at rest the lever is central between the banking pins. If not, adjust the position of the hair spring until it sits in a central position.

Note on screws

In high-quality watches you might come across screws with one, two, three or more small dots punched into the heads. These are matched by dots punched into the movement next to the screw holes. When you reassemble the watch, make sure that the dots are correctly matched.

Winding up

At this point you are ready to wind the watch. A bit of care is required as the fusee chain will not necessarily be aligned with the grooves on the fusee. Fit a key to the winding arbor and slowly wind in an anticlockwise direction. Keep an eye on the chain and make sure that it is feeding into the grooves correctly. If not, use tweezers to move it up or down the spring barrel to realign it. As the chain comes to the top of the fusee, it should push the fusee stop work up against the bottom plate; this then meets the stop piece on the fusee and should prevent any further rotation. Keep a very close eye on the stop work as you reach the end to make sure that it functions correctly. If you reach the end of the chain before the stop work has engaged, do not wind any further (stop before the chain is pulled off the spring barrel, as this could damage it). If it does not engage, you will need to dismantle the movement and adjust the stop work.

If all is well, a shake of the watch should set it going. Put it aside in a safe place and leave it for twenty-four hours; if it is still running, then you can fit the dial.

After a little while, you will develop a sense as to whether a particular watch will run well or stop soon. You will know from the way it ticks and how the balance swings.

Fitting the dial

Fit the motion work and dial washer, then fit the dial directly on the movement and fix it with brass pins.

Fitting the hands

If the hands are of blue steel, they only need to be carefully wiped with a clean piece of kitchen towel and fitted. If they are gold, they may need to be cleaned with metal polish. Lay them on a flat surface and wipe along the length with a piece of kitchen towel soaked in polish. Do not wipe towards the centre mounting as it is very easy to catch the point and bend the hand. Finish by cleaning with a fresh piece of kitchen towel.

Once cleaned, fit the hour hand, pressing it down with a pair of tweezers until it is almost flush with the dial; make sure it is not in contact at any point. If it is, this may cause the watch to stop and can also leave marks on the dial. Adjust the hour hand so that it is pointing to 12 o'clock. Fit the minute hand (aligned with 12 o'clock), and press it down with a pair of tweezers. Now fit the second hand and press it down (not too much or it can catch on the tube). Using the key, rotate the hands through twelve hours to make sure they do not catch (always move them in a clockwise direction to prevent possible damage to the movement).

Regulation

Once the watch is working, you will need to bring it to 'regulation'. This involves adjusting the regulator arm backwards and forwards until the watch keeps good time. The regulator arm usually moves over a scale engraved on the bottom plate (usually marked fast and slow). As the arm is moved, this in turn moves the curb pins along the length of the hair spring, so shortening or lengthening its effective length. The effect of changing the length of the hair spring is to change the rate of rotation of the balance (the beat); the longer the hair spring, the slower the beat, the shorter the spring the faster the beat.

Things that might stop the watch from working

Like all mechanical devices, watches do not always work as intended. This may be due to a number of reasons. It may have a damaged or badly fitted part, there may be too much friction in the train or there may be any number of other faults.

Do not assume that all the parts found in a watch started life with that watch. It is quite possible that someone has in the past replaced a damaged or missing part in order to try to get it working, which may not have been successful. The replacement part may therefore be the cause of the movement not working.

If the balance does not swing freely, this can be caused by a number of problems.

- Check that the hair spring is level and not catching on the plate holding the lever and escape wheel or balance cock.
- Check that the pivot jewels are clean and not damaged.
- If the balance moves sluggishly, you may need to burnish the top and bottom pivot with a Jacot tool.
- If the safety pin on the lever is bent forward and rubbing on the table roller, dismantle and remove the lever, then carefully bend the pin backwards, reassemble the movement and test it.

If when the balance is shaken, the lever does not remain engaged with the safety roller, it is usually because the guard pin which passes through the crescent on the roller is bent so that it can 'jump' out of sync with the balance. To fix this, remove the lever, then carefully bend the safety pin forwards a small amount. Reassemble and test it. As the balance swings from side to side the lever fork should engage with the impulse pin. The safety pin should pass through the crescent cut in the roller and at no time should it rub on the roller.

Another problem is a lack of power to the lever. Remove the balance and balance cock then, using a piece of peg wood, push the lever from side to side. If it is moved a small amount, it should snap back when you let go. If you move it a bit more, it should snap over to the other side. If this does not happen, the problem lies in the train. Let down the tension in the spring and remove the lever, then wind it up a little; the escape wheel should rotate freely. Do not wind it up too much, as it could damage the wheel pinions if they are allowed to run too fast. If it does not turn freely check that there are no bits trapped in the leaves of the pinions or between the teeth on the wheels.

Another cause of a sluggish train may be a distorted fixing plate. Check that both the top and bottom plates are flat. If not they need to be flattened; you can straighten them by applying a small amount of pressure in the opposite direction. Reassemble and check if the train is free-running.

THE ENGLISH THREE-QUARTER-PLATE AND OTHER MOVEMENTS

A less common, but still quite popular style of movement is the English three-quarter plate. This is really a variation on the fusee centre chronograph, so it should be dealt with in much the same way. The main differences are that there is not usually a centre second hand or a stop lever.

Other movements

- Rack lever: this is very similar to the table roller lever except that the end of the lever has a multi-toothed rack instead of the usual fork, which engages with a pinion on the balance.
- Early English cylinder: clean and repair as for the English lever but *do not* remove the balance until the tension in the spring has been let down.
- Massey lever: there are five types numbered I, II, III, IV, V. All are variations on the early lever and as far as repair is concerned, treat them like the English table roller lever.
- Savage two pin lever: again this is very similar to the table roller lever and needs no special attention.

THE ENGLISH FUSEE VERGE

The verge pocket watch developed during the sixteenth century from small clocks, but it was not until the late seventeenth century, with the invention of the hair spring by Robert Hook, that it became a viable pocket watch that could maintain reasonable accuracy. They are usually found in silver, gold or gilt pair cases, but later ones may be found in double-bottom cases similar to the English fusee lever.

I describe here a typical English fusee verge from the late eighteenth and nineteenth century.

Figure 73 English pair case fusee verge

Removing the movement from the case

Figure 74 Disassembled inner case

This type of watch most commonly comes in a pair case. Later models may be cased in double-bottom ones. The pair case consists of an inner and outer cover. First remove the outer case by pressing the button on the side and opening the front. The inner case can then be lifted out using the pendant. The inner case is opened by inserting a case knife between the front and back at 6 o'clock and levering the front up, then using a pin pusher or a piece of steel wire just slightly smaller in diameter than the case pin to push it through. In some cases, you may need to tap the end of the wire with a hammer in order to push the pin out. Once this has been removed, the bezel and movement can be lifted away from the back of the case. For other case styles, see details for the English fusee lever (page 69).

Removing the hands

Having successfully removed the movement, you can now remove the hour and minute hands. Do not apply too much pressure to the front of the movement as this could damage the balance, unless it is fitted with a dust cap. Use a purpose-made hand remover which is designed to lever the hands up without damaging the dial. Press the button on top of the tool, which will cause the jaws to open. Press it down over the hands (the centre section is spring loaded) and release the button. The jaws will close under the rim of the hands. Press the whole handle down, which will cause the jaws to lever up the hands and free them from the cannon pinion. Be careful not to bend the hands; position the hand remover so that it does not put any strain on them. It is uncommon for a verge to have a second hand but if one is fitted it can be removed by holding a craft knife near the stem and levering it up; hold your finger over it, so that it does not fly off.

Figure 75 Removing hands

Removing the dust cover

Figure 76 Rear of movement showing dust cap (note spring clip is missing)

Most watches of this type do not have a dust cover, but occasionally you will come across one that does (see figure 76). To remove it push the curved bar using the small nib projecting from the middle; it should slide round and release the cover, which can then be lifted clear.

Removing the dial

The most common method of retaining a dial on a fusee verge watch is via tapered brass pins (usually three) which pass through holes in the dial feet. These should be spaced evenly round the dial. To remove the pins, either pull them out with a strong pair of tweezers or, if this proves to be difficult owing to the tight fit of the pin, get hold of the end with a pair of cutters (only apply enough pressure to grip it and not cut through) then pull back on the handles against the edge of the bottom plate. This should lever the pin out. Be aware that sometimes the dial feet may have become bent. If this is the case, you may need to lift the dial on one side first before you can remove it completely. Be very careful about bending the dial feet, as this may cause damage to the front of the dial enamel; it is much

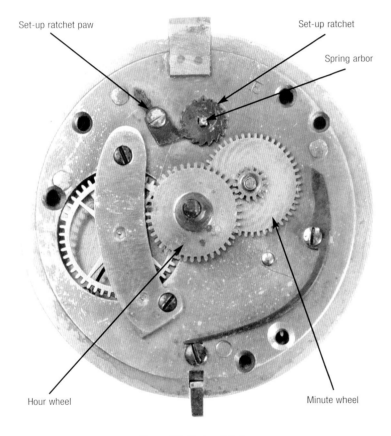

Figure 77 Under-dial view

better to leave them as they are unless this prevents you from removing the dial.

In some watches the dial is fixed, using brass pins, to a dial plate, which in turn is pinned to the top plate. This is removed in exactly the same way as above.

With tweezers, remove the motion work and, if present, the dial washer.

Cleaning the dial

If the dial is undamaged, it only needs to be cleaned in soap and warm water (not too hot, otherwise this may damage the enamel). After cleaning, dry it off with a paper towel. If the dial has hairline cracks, these can sometimes be made to look less obvious by soaking in a mild bathroom cleaner (Bath Power by OzKleen seems to work well). It is normal for the numbers to be an integral part of the enamel dial and they are therefore unlikely to be removed by any cleaning. However if there is other writing on the dial (maker's name etc.) care must

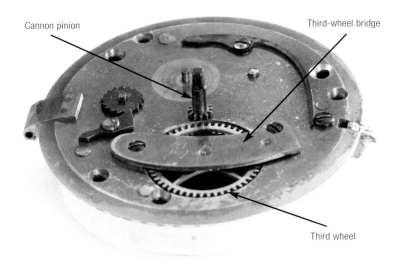

Figure 78 View of cannon pinion

be taken to make sure that this is not removed during cleaning, so check a small area first. Once cleaned, put the dial
safely to one side.

Removing the cannon pinion

The cannon pinion is held in place by a friction fit. To remove it
you need to take hold of it with a pair of small pliers and, with
a twisting action, pull it firmly away from the front of the watch.
Be careful not to apply too much force as you may distort it. If
it proves to be very stiff, a small amount of oil should be applied
and left overnight to work its way in.

Letting down the main spring

Before removing any further parts, it is *very* important that you
release the tension in the main spring. If there is no tension in the
spring, it may be due to a broken spring or a broken, loose or
missing fusee chain. If this is the case, you can skip the next
section.

Looking down on the top plate, where the dial was, you
should see the third-wheel bridge held in place by two screws.
While keeping pressure on the bridge, undo and remove both
screws. It is very important that you do not allow the bridge to
detach from the plate. Fit the correct-sized key to the winding
square and hold it firmly with your thumb and forefinger. Then
carefully lever off the bridge and remove the third wheel. You
should find that the key now wants to turn; allow it to do so, all
the while keeping it firmly between your fingers until the spring
has completely run down. Then remove the key.

Slacken off the screw that holds the set-up paw against the
set-up ratchet and push the paw away; this should release the
last bit of tension in the spring. You can now pull the fusee chain
out so that it unwinds from the spring barrel and fusee. Carefully
unhook the chain from the spring barrel, and pull it clear. Then

unhook the other end from the fusee. Remove the ratchet wheel by pulling it clear of the spring barrel arbor.

In older verge watches (pre 1770s) you may come across a different set-up arrangement on the fusee. This consists of a wheel attached to the underside of the spring barrel and connected to the central arbor, which engages with a worm whose end is at the edge of the bottom plate, which can be turned via a key. To let down the tension in the spring, fit a key to the square on the end of the worm and wind it anticlockwise until the fusee chain goes slack. The watch is now safe to dismantle. Note that in this design there is not usually a separate bridge covering the third wheel or spring barrel.

Removing the fourth (contrate) wheel

Figure 79 View of top plate with third wheel removed

This step may not be possible on older watches. Using a pair of tweezers, lift the fourth wheel up a little so the lower pivot disengages from its pivot hole, then manoeuvre it so that the rim of the wheel can be lifted clear of the edge of the centre wheel (you will probably need to angle the fourth wheel). It should then be possible to lift it clear of the movement, but do not pull it clear if there is any resistance as this may damage it. If you cannot get it free, remove it later once the plates have been separated.

Removing the balance

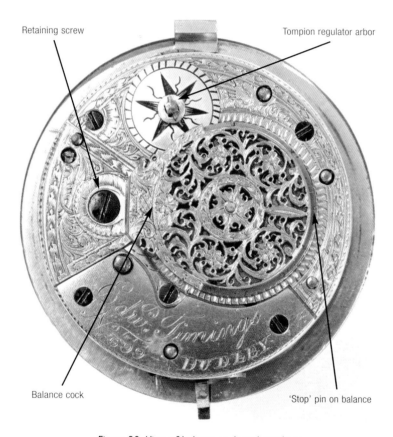

Retaining screw

Tompion regulator arbor

Balance cock

'Stop' pin on balance

Figure 80 View of balance cock and regulator

The balance is held in place by the balance cock, which in turn is fixed in place by a single screw. Hold the movement face down and level, then undo the screw and remove it. Using a strong pair of tweezers, carefully lift the balance cock clear of the balance. Remove the brass pin that fixes the hair spring to the hair-spring stud, and then feed the hair spring through until it is free. Lift the complete balance assembly away from the bottom plate, making sure that it is free of the curb pins on the regulator before you lift it up.

Now remove the regulator. This is usually a separate plate attached to the bottom plate and held in place by screws. Remove

Hair-spring mounting stud and retaining pin

Curb pins

Figure 81 View of movement with balance cock removed

Hair spring

Pallets

Figure 82 Verge balance

Figure 83 Underside view of regulator plate showing rack

these and lift the assembly clear. If the regulator is of the Bosley type then the assembly should come away completely. If it is a Tompion type, there may be parts left behind when you lift the assembly; remove these as well.

Now you can remove the spring barrel bridge. Unscrew the two screws that hold the spring barrel bridge in place, and then lift the plate clear. Now lift the spring barrel away from the movement.

Note that on a large number of verge watches, including the one illustrated here, there is no spring barrel bridge. This means that the spring barrel can only be removed once the bottom plate has been removed.

Dismantling the plates

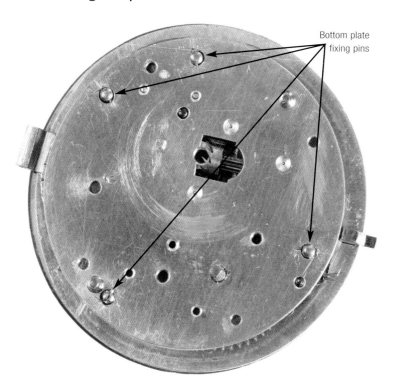

Bottom plate
fixing pins

Figure 84 View of bottom plate showing fixing pins

Almost all English fusee verges that you are likely to come across are held together by brass pins. To remove the pins, push them through with a pair of pliers by placing one jaw against the thinner end of the pin and the other against the pillar (on the other side) above the pin (the jaws will therefore be at a slight angle). Press the jaws together and the pin should pop out. The ones under the balance cock or spring barrel cover will be only just longer than the width of the pillar head. To remove these, try the same method as for the longer ones above. If this fails, try removing them with a piece of steel wire that has been filed

Figure 85 Bottom-plate assembly

sharp at one end and hardened. By applying pressure to one end of the pin, it should be possible to push the pin out (a light tap with a hammer may be helpful).

Once all the pins have been removed, you can remove the bottom plate. You will need to lift the plate carefully clear of the four pillars. Having separated the plates you can now remove the centre and contrate wheels, then remove the fusee and, if still present, the spring barrel.

Removing the crown wheel

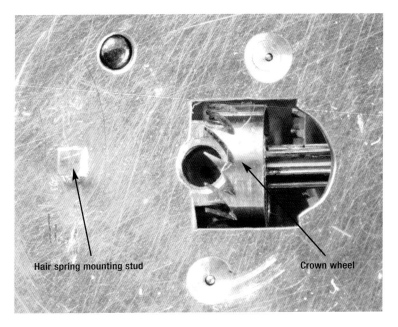

Figure 86 Detailed view of bottom plate showing crown wheel

The crown wheel is mounted on the underside of the bottom plate between two pivots, one of which consists of a peg held in a brass mount. This pivot is blind and its position can be adjusted by pulling the plug in and out of the mount. The other end is held by the potance plate which also carries the lower pivot of the balance. This is fixed by a single screw. Remove this

135

Figure 87 Underside view of bottom plate showing fusee stop work

and very carefully lever it away from the bottom plate. It is very important that you are careful as you might very easily break one of the pinions on the crown wheel.

Cleaning the plates

Before cleaning the plates, it might be worth removing the fusee stop work spring, as this can easily be bent or damaged by cleaning. It is held in place by a single screw; undo this and lift the spring away from under the fusee stop.

The best way I have found to clean the plates and other parts is to use a metal polish. Some older books on watch repairing advocate the use of benzine to remove oil and grease, but this is not only difficult to obtain but is also a poison and should not be used. Using kitchen towel and metal polish, rub the surface until all the dirt, marks and grease have been removed. After polishing the plates with a clean towel, the next step is to clean the pivot holes. This is done using a length of peg wood (see list of suppliers in appendix A), whose end has been sharpened to a point with a craft knife. Insert the sharpened peg wood into each pivot hole and rotate it a few times. Repeat until the wood comes out clean. Do not be tempted to reuse the peg wood on another pivot hole without resharpening it, as this can lead to the point breaking off and sticking in the hole. If this happens, resharpen the end and try to push the broken bit out from the opposite side. When you are happy that all the pivot holes are clean, return the parts to the tray.

Do not forget to clean the third-wheel bridge as well (if fitted) and the pivot holes of the crown wheel. Be careful of the blind pivot hole, as it is very difficult to remove a broken piece of peg wood from this (it may be done with the pointed end of a fine burnisher). Also clean the top and bottom pivot holes of the balance. Note that the top pivot in the balance cock can often be blind, so take care.

If the pivot holes are straight into the brass plates (i.e. not jewelled), which is most likely with verges, check that they are still round by examining them with a loupe. If they are not, it may be necessary to close up the hole with a staking tool or, if they are very worn, rebush them (see page 157).

Cleaning the wheels and pinions

Examine each wheel to make sure that it is flat and that the pivots are straight. If they are, clean both pivots by holding a piece of peg wood so that the pivot sinks into the blunt end and rotate it a few times. If the pinion vanes have dirt between them, this needs to be cleaned out. Sharpen a piece of peg wood to a flat point and push it along each of the vanes until all the dirt has been removed. A rub with an old (clean) tooth brush can remove most general dirt. Do not use too much force as this can distort the wheel.

Fourth (contrate) wheel

Third wheel

Centre wheel

Figure 88 Centre, third and fourth wheels

Checking the spring

One of the more common problems with pocket watches is that the spring is often broken. To check this, remove the spring barrel

Figure 89 Spring barrel with cover removed

cover by inserting a screwdriver into the slot and carefully levering it up. Check that the spring is unbroken and still connected at both ends. Either the outer end should have a hole which hooks over a pin on the inside of the spring barrel or it should have an oblong piece of steel riveted to the end of the spring, which then fits a matching hole in the barrel. The inner end should fix onto a hook on the barrel arbor. If this all looks to be all right, then you only need to apply a small amount of oil and replace the cover. This fits into a small groove and may need to be pressed back into place with a pair of brass-faced pliers. It should be flush with the barrel when properly fitted.

Checking the fusee

Figure 90 Underside of fusee

The fusee consists of a cone-shaped block of brass with a spiral groove cut into it, along which the fusee chain winds. Even if it appears to be working correctly, it is good practice to dismantle and clean it. The fusee consists of two main parts. The body is the fusee cone (made of brass), which is attached to the steel winding square. Below this is the ratchet. Both parts are held together by a pin which passes through the lower part. To dismantle it, push out the pin and remove the collar. Next remove the wheel and ratchet.

First examine the ratchet wheel on the underside of the fusee cone for wear. If it looks all right, proceed to the ratchet. This consists of a paw that engages with the ratchet wheel and is held against it by a spring (which is an integral part). The paw is riveted into this and is free to move. If the watch has been badly treated, this may be damaged or bent due to the fusee being forced backwards. A replacement paw can be made but, because of its small size, this is quite difficult. It may be better to try to replace a damaged ratchet with one from a scrap fusee. The same is true if the ratchet wheel is badly damaged.

Figure 91 Disassembled fusee

If all is well with the fusee parts then just give the inside a clean and oil then reassemble the parts. Finally reinsert the pin to hold it together.

Fusee chain

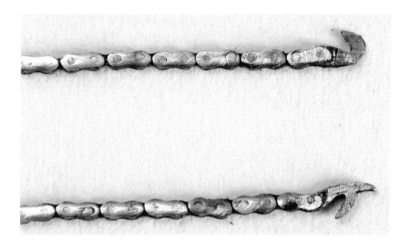

Figure 92 Fusee chain

The fusee chain consists of dozens of links all hand riveted together to form a very strong and flexible chain with a hook at either end. This is designed to connect the fusee and the spring barrel. When the watch is wound up, it is transferred from the spring barrel onto the fusee; as the watch runs down, the chain winds back onto the spring barrel. If the chain and both the hooks are good, then it only needs light oil. If the chain is stiff and discoloured, it needs a little more attention. Take a length of peg wood about 5mm in diameter and cut a slot in one end. Hold upright with the chain held across the slot. Place a few drops of oil on the chain, then pull it backwards and forwards through the slot, moving along its entire length. Do this until the chain feels reasonably free. Wipe the excess oil from the chain.

If the chain has broken (a very common cause of a fusee watch not running), then it needs to be either replaced or repaired (see page 163). When choosing a replacement, it is important that it is of similar thickness and length. If the replacement chain is thicker than the original one, it may not fit the groove cut in the fusee; if it is too thin, there is the possibility that it will not be strong enough for the spring and subsequently break due to the strain. If the original chain is missing, then you can check whether a replacement one is the correct length by fixing the hook at the start of the fusee and winding it onto the fusee until you reach the top (just below the 'stop'). There should be a minimum of 3cm of the chain left to attach onto the spring barrel. A bit more is not a problem, but any less will probably lead to the chain breaking when the watch is first wound up due to the stop not engaging.

Before you reassemble the watch, make sure you fix the stop work spring in place. The flattened end should go under the stop work and, when screwed back in place, should lift it clear of the surface of the bottom plate. If you press down and let go, the stop work should spring up again. It is designed to be pushed flush with the bottom plate when the fusee chain reaches the last turn on the fusee. The end of the stop work should then engage with the stop on the fusee and prevent it from turning any more.

Reassembling the watch

Assuming that there are no parts missing and any necessary repairs have been made, it is now time to reassemble the watch. Take the bottom plate and fit one end of the crown wheel into the adjustable pivot. Fit the other end into the potance pivot and re-engage the positioning pins in the bottom plate. Push the potance flush with the bottom plate, and then spin the crown wheel to make sure it is correctly placed. If it does not run freely, use a screwdriver to adjust the position on the pivot (see figure 87) until the crown wheel spins freely. Once it is running freely, screw the potance in place. The crown-wheel pivot nearest to the balance is held by a pivot hole in a small piece of brass that can be moved from side to side. This adjustment allows the position of the crown wheel to be moved relative to the balance pallets. You should aim for the crown-wheel pivot to be in line with the balance staff. Only adjust this if the watch was not running before you dismantled it. Apply a small amount of oil to both pivots and to the teeth of the crown wheel.

Continue to hold the bottom plate upside down, and place the fourth wheel in position (if there is no spring barrel bridge, then fit the spring barrel as well). Put the fusee in place and then finally the centre wheel. Bring the top plate carefully down over the bottom plate, allowing the centre-wheel pivot to pass through the appropriate hole, then make sure the two plates are aligned (the pillars lined up with their corresponding holes). Slowly bring the two plates together so that the pillar heads start to go through the holes. At this point check that everything is still in place and push them together until the fourth wheel touches the top plate. Using a pair of tweezers, move the fourth wheel until the pivot drops into its appropriate pivot hole. Push the bottom plate down until it is in the correct position. Check carefully that no parts are trapped and everything is in the correct position. Pin the plate in place with the four brass pins (remember to refit the short ones in the correct positions). Turn the movement over and, if it has a bridge, fit the third wheel and the bridge. Make sure the fourth-wheel pivot is correctly aligned before screwing the bridge in place. Now turn the fusee anti-clockwise with a key; the train should run freely. Do not put too much pressure on the fusee as this might lead to the train running too fast and damaging the pivots, especially the crown wheel.

Fitting the spring barrel

If it has not already been fitted, place the spring barrel with the ratchet arbor through the top plate. Fit the spring barrel bridge and fix it in place with the two screws.

Fitting the regulator

The regulator is usually held in place by three or four screws. Position it over the mounting holes (if the regulator is of the

Tompion design, place the rack and dial in place first) and fix it in place with screws. Check that the lever on a Bosley regulator moves freely, or that the disc on a Tompion regulator turns freely.

Fitting the fusee chain

Rotate the fusee both ways with a key and check that it moves correctly. Leave the fusee so that the attachment point for the fusee chain is on the outside edge. Take the fusee chain and, holding the movement edge on, drop the chain past the spring barrel behind the pillar so that the end comes out by the fusee (note that a fusee chain has a different type of hook at either end; the pointed one fits on the spring barrel). Using tweezers, hook the end of the fusee chain onto the fusee; there should be a slot cut into the fusee, and in the middle of this there is a steel pin, to which the hook must be attached. Pull the chain until the fusee stops rotating.

Using a key attached to the set-up arbor, rotate the spring barrel anticlockwise until the small slot for the fusee chain is on the outside. Hook the free end of the fusee chain into the slot. While keeping the chain away from the movement with your finger, rotate the spring barrel with the key and feed the chain onto it (try to keep the chain parallel and away from the bottom of the spring barrel). When all the chain has been wound onto the barrel, keep pressure on the spring barrel, remove the key and fit the set-up ratchet over the arbor and engage the paw. Then using a screwdriver, push the ratchet round a few notches keeping the paw engaged, then tighten the paw screw to hold it in place (be very careful not to slip). The fusee chain should be under tension.

Fitting the balance

Before the balance is fitted, oil the pivots, including the top and bottom balance pivots; do not over-oil as this can lead to a build-up of dirt. Use an oil pin (a tool specially designed for the job) and transfer a drop to each pivot; aim to apply just enough oil to partly fill the indentation around the pivot.

Place the balance on the bottom pivot and, using tweezers, feed the end of the hair spring through the hole in the hair spring stud. Make sure that the outer turn of the hair spring passes through the index curb pins. Push it through until the balance is correctly aligned, i.e. the two pallets sit at an equal angle either side of the centre line passing through the crown-wheel pivot. On some verges there is a pin sticking up from the rim of the balance wheel (see figure 80), which should be positioned so that it is central to the balance cock. This provides a stop if the balance oscillates too much. Fix the hair spring in place with a brass pin, and then fit the balance cock.

Give the movement a light shake so that the balance swings; check that it moves freely and that when it stops the pallets are equally spaced either side of the crown wheel. If they are not, you will need to adjust the hair spring by loosening the brass pin and moving the hair spring until the pallets are central. Once everything is set up, you can screw the balance cock in place. It is a good idea to give the movement a gentle shake so that the balance is swinging, then slowly tighten the balance cock screw until it is fully home. Check that the balance still swings freely; try it in several positions (face up, face down etc.).

Note on screws

In high-quality watches you might come across screws with one, two, three or more small dots punched into the heads. These are matched by dots punched into the movement next to the screw holes. When you reassemble the watch, make sure that the dots correctly matched.

Winding up

At this point you are ready to wind the watch. A bit of care is required as the fusee chain will not necessarily be aligned with the grooves on the fusee. Fit a key over the winding arbor and slowly turn it in an anticlockwise direction. Keep an eye on the chain and make sure that it is feeding into the grooves correctly. If not, use tweezers to move the chain up or down the spring barrel to realign it. As the chain comes to the top of the fusee, it should push the fusee stop work up against the bottom plate; this then meets the stop piece on the fusee and should prevent any further rotation. Keep a very close eye on the stop work as you reach the end to make sure that it functions correctly. If you reach the end of the chain before the stop work has engaged, do not wind any further (stop before the chain is pulled off the spring barrel as this will almost certainly damage it). If it does not engage, you will need to dismantle the movement and adjust the stop work. Putting a small screwdriver under the stop piece and pushing down on the end should lift the centre section and therefore allow the stop work to engage sooner. Reassemble the movement and try again.

If all goes well, a shake of the watch should set it going. Put it aside in a safe place and leave it for twenty-four hours; if it is still running, then you can fit the dial.

After a little while, you will develop a sense as to whether a particular watch will run well or stop soon. You will know from the way it ticks and how the balance swings.

Fitting the cannon pinion

This is held in place by friction. Place it over the centre wheel pinion and press it down with pliers. You should be able to rotate it forwards and backwards but there should be some friction or it will not rotate when the centre wheel moves.

Fitting the dial

Fit the motion work and dial washer. Then, if the dial is fitted to a dial plate, first fit this to the dial and fix it in place with brass pins. Then fit the dial and plate to the movement. If not, then fit the dial directly on the movement. In both cases, fix it with brass pins.

Fitting the hands

If the hands are of blue steel, they only need to be carefully wiped with a clean piece of kitchen towel and fitted. If they are gold, they may need to be cleaned with metal polish. Lay the hands on a flat piece of newspaper and wipe along the length with a piece of kitchen towel soaked in polish. Do not wipe towards the centre mounting as it is very easy to catch the point and bend the hand; be particularly careful if there are any complex bits. Finish by cleaning with a fresh piece of kitchen towel.

Once cleaned, adjust the minute hand arbor so the square allows the hand to point to 12 o'clock. Fit the hour hand (aligned with 12 o'clock), and press it down with a pair of tweezers; it should be just clear of the dial surface. Now fit the minute hand and press it down with the correct-sized key. It should not touch the hour hand. Using the key, rotate the hands through twelve hours to make sure they do not catch (always move them in a clockwise direction to prevent possible damage to the movement).

Regulation

Once the watch is working, you will need to bring it to 'regulation'. This involves adjusting the regulator arm backwards and forwards until the watch keeps good time. The regulator arm usually moves over a scale (usually marked fast and slow) engraved on the bottom plate if it is a Bosley type or on an engraved disc if it is a Tompion type. As the arm or disc is moved, this in turn moves the curb pins along the length of the hair spring, so shortening or lengthening its effective length. The effect of changing the length of the hair spring is to change the rate of rotation of the balance (the beat); the longer the hair spring, the slower the beat, the shorter the spring the faster the beat.

Things that might stop the watch from working

Like all mechanical devices, watches do not always work as intended. This may be due to a number of reasons. It may have a damaged or badly fitted part, there may be too much friction in the train, or there may be any number of other faults.

Do not assume that all the parts found in a watch movement started life with that watch. It is quite possible that someone has in the past replaced a damaged or missing part in order to try to get the watch to work. This may not have been successful. The replacement part may therefore be the cause of the movement not working.

If the balance does not swing freely, this can be caused by a number of problems.

- Check that the hair spring is level and not catching on the bottom plate or balance cock.
- Make sure that the hair spring is symmetrical.
- Check that the pivot holes are clean and not damaged.
- Are the balance arms catching on the hair spring stud? If so, you can carefully reduce the height with an Arkansas stone.
- Are the pivots smooth? If not burnish them with a Jacot tool.

If the watch runs but the balance stops repeatedly, it may be due to poor engagement of the crown-wheel teeth with the balance pallets. Check that all the crown-wheel teeth are the same length, and are even. If they are not you may be able to adjust them using a fine file, but be careful not to remove too much metal. Fit the crown wheel in some form of mount (Jacot tool or lathe) so that it can be rotated, and even out the points with a file (be careful not to apply too much pressure as you might damage the pivots). Then remove the crown wheel and file the teeth so that they are all even. Remount and adjust the depth of the teeth with respect to the pallets.

Another problem is that as you change the angle of the movement, the crown wheel disengages from the pallets and the train runs at high speed. This can be due to a loose crown wheel. The crown wheel can move backwards and forwards between the pivots. To fix this, carefully move the adjustable crown-wheel rear pivot so that it is closer to the centre of the movement. The rear pivot consists of a small brass tube with a flattened end which is mounted on the underside of the bottom plate (see figure 87). The tube is a friction fit and can be moved backwards and forwards. To adjust it, take hold of the flat part with a pair of pliers and, with a twisting action, push it towards the centre. Be very careful that you do not push it too hard and damage the crown-wheel pivot. This should bring the crown-wheel teeth closer to the balance pallets. If this does not solve the problem, you may need to move the front pivot closer to the balance. Remove the small brass piece and place it over a hole on the staking tool table then, using a round-ended stake, lightly tap it so that the centre is pushed in slightly. You may need to adjust the pivot hole with a burnisher as it might have closed up. Refit the crown wheel and adjust the rear end pivot as above, so the crown wheel rotates freely but does not move backwards and forwards. Reassemble the movement and check whether this cures the problem.

French/Swiss fusee verge

The eighteenth-century design of the French/Swiss fusee verge is similar to that of the English, but differs in one or two areas. The balance cock is usually circular with two mounting lugs held in place by two screws. And the design of the balance potance and crown wheel pivot is more complex.

The process of dismantling one of these watches is very similar to that of an English fusee verge. Assuming that the watch was ticking before it was dismantled, there is no need to adjust the potance or crown-wheel pivot.

Closer examination of these two devices will show that one is designed to allow the position of the centre of the crown-wheel pivot to be moved relative to the balance staff. The other adjustment that can be made is to the depth of engagement of the crown-wheel teeth with the balance pallets. In making these adjustments, the aim is first to set the pivot of the crown wheel in line with the balance staff. The depth of the crown-wheel teeth can only really be adjusted by trial and error. Once the watch is going, turn the adjustment screw right and left while listening to the movement and when the 'tick' sounds at its strongest you have hopefully reached the optimum position.

THE CASE

Cleaning

If the case is silver, then use a silver cleaner to remove all the oxide and marks. There are several suitable cleaners available (paste, cream liquid and impregnated cloth); I find that Goddards long-term silver polish and Silvo Duraglit both work well. After polishing with kitchen towel, check the general state of the case. If there are dents, they can sometimes be removed by a round-faced brass hammer. Place the case on a thick layer of newspaper and then tap from the inside until the dent has been reduced. Before doing this, it is worth considering whether it is better to leave the dent alone, as it is possible to make things worse if you are not sure what you are doing; it is probably worth practising on an old scrap case first. If the case is in good condition, then the only other thing to do is to apply a small amount of oil to the hinges.

Gold and gold-plated cases should be cleaned with a brass cleaner. With gold cases, it is also possible to remove dents, as explained above for silver ones. When it comes to gold-plated cases, it is more difficult to remove dents as the base metal is usually brass and therefore a bit harder than gold or silver; furthermore the gold plating is very thin, so it can easily be damaged.

Repairing

Case repairs are difficult as they often require additional skills similar to those in jewellery work. It might be better either to leave any fault alone (if it does not affect the use of the watch), or to pass the work over to a jeweller. Some things can be repaired; it is not too difficult to replace a case spring. You will need to remove the broken one; if you are lucky it will be held in place with a screw. If this is the case then unscrew it and remove the spring. You will then need to obtain another one from a scrap case. It must be of the same size and shape or it will not work. If you are unlucky, the spring will be held in place with a steel pin. The only way I have found of removing these is to drill the pin out (make sure that you stop before you go through the case on the other side). Remove the spring and replace it with the new one. Then using a piece of steel wire of a suitable diameter (it should be a tight fit), tap it into the hole. Cut off the excess and carefully file it flat (try to avoid damaging the case surface).

On cases with integral winders, a common problem is that the winder will not 'click' into one of two positions (in to wind, out to set the hands). This is usually due to damage to the tube that holds it in place. To fix this you need first to remove the winder. Hold the square section of the winder (the part that engages with the watch movement) with a pair of pliers and then twist the winding crown anticlockwise with your fingers. The crown should unscrew and the square section can then be pulled out. If the crown will not unscrew, apply a small amount of oil and leave it to soak overnight. The winder is held in the two positions by an inset that is screwed into the pendant. This has four spring leaves that grip the winder. Quite often one or more of these will have broken off; if this is the case then you need to replace it. Use a screwdriver that is wide enough to fit the slots on both sides, and unscrew it. There is little that can be done to repair the old one, so a replacement must be obtained from a scrap case or from old stock. It should be screwed into the pendant to the same level as the original one.

Another common problem is damage to the hinges. Often the pin that holds the hinges together wears through or snaps. In order to replace this, you first need to remove the broken parts. This is not easy as the ends will have been smoothed over. If the two parts have become separated, it is easier to push the pieces out from the inside. Using a length of steel wire just smaller than the inner diameter of the hinge, gently tap the broken bits out. Be aware that, particularly with silver cases, too much force may cause the tubes that form the hinge to separate from the rest of the case. Put the two parts together and, using a reamer, lightly twist it to clean out the tubes (do not remove too much metal). Then take a length of tapered nickel or brass wire and push it into the hinge until it fits tightly. Mark the points where it enters and exits the hinge. Pull it out and cut the wire to the correct length then file the ends smooth. Refit the wire into the hinge and push it into place with a piece of peg wood, checking that there are no sharp ends. If the case is gold or silver, you can very carefully file the ends flush with the case. Only do this with a very fine file, then finish off with '000' grade steel wool and finally with metal polish. If the case is gold plated, it is not a good idea to file it, as the very thin layer of gold can easily be removed.

Most machine-made cases from the early twentieth century are made to a standard size and are interchangeable. It is therefore possible to remove a watch movement from a damaged case and put it into a good one. This is not really possible with pre-twentieth century watches as the cases were usually made for a specific movement.

Crystals

If the watch still has its original glass crystal, it is best not to try and remove it unless it is badly scratched, as it is possible that it may become damaged around the edge. If it is badly damaged or missing, you will need to replace it, either with a plastic one (available from watch-repair material suppliers) or, much better, by a glass one. The latter can still be bought from auctions etc. (usually only as a number of glasses sold as a lot). There are also some people who will sell just one specific crystal, if you supply the required diameter. In an ideal world the crystal should be chosen so that it 'pops' into place and does not require any glue. In practice you will not always have the exact size, and therefore it may need to be glued in place. Use a 'soft' glue that can easily be removed later if necessary. Once it has dried, remove any excess with a piece of peg wood so as not to damage the case finish.

If you only have a crystal that is a little larger than the required size, it is sometimes possible to grind the edge down to fit, using a stone designed to sharpen woodworking tools. The best types are those made of synthetic diamonds set in a metal and plastic matrix. Make sure that the surface is kept wet with water. Holding the crystal between thumb and forefinger, slowly rub backwards and forwards along the edge and at the same time rotate the crystal to make sure that it remains round. Unless you are very careful, the edge can easily become chipped, so take your time and do not apply too much pressure. Before trying the crystal in the bezel, make a few rotations on the stone at 45 degrees in order to round off the bottom edge. It is probably better to make the crystal slightly smaller that the bezel and glue in place rather than trying to make it a tight fit and pop it into place, as this may lead to chipping around the edge.

When choosing a suitable crystal, make sure that it is not only the correct diameter but also high enough to accommodate the hands and cannon pinion. There should be sufficient space to allow the hands to move freely. As you close the front, make sure that the end of the minute hand is not pressed down as this can cause it to catch on the hour hand and stop the watch.

PROBLEMS THAT CAN AFFECT MOST WATCHES

The watch will not run after cleaning/repair

The following is a list of the most common causes.

- The watch stops with both hands in line, usually caused by one of the hands catching on the other. Check from the side with a loupe and lift the hand with tweezers if it is catching. It can also be due to the hour hand catching on the second hand. Check also that the minute hand is not in contact with the crystal.
- The balance catches on the bottom of the case or dust cap. Over years of use, the bottom of the case may have become distorted and therefore interfere with the balance. It may be possible to correct this by pressing the bottom outwards with your thumbs.
- The watch only works face down. The balance may be catching on the underside of the balance cock. Check with a loupe, you may need to adjust the shape of the balance cock to prevent this (use brass-faced pliers to bend it). Also check whether any of the jewel-mounting screws on the balance cock are touching the balance. If they are, remove and file down the end a bit.
- The watch only works face up. The balance may be catching on the hair-spring stud. If so, you may be able to file a little metal off the top, but be careful not to file too much as you may cut through the hair-spring hole.

- There may be too much movement in the balance. Using a pair of tweezers gently try to move the balance up and down. There should be a little play but excessive movement can mean that the impulse pin and table roller are misaligned when the watch is turned over. This problem can also be fixed by adjusting the shape of the balance cock.

- The balance may be too tight. This may be the case if the balance cock has become distorted and is pressing down on the pivots. To check whether this is the problem, slightly undo the screw holding the balance cock in place. If the balance now swings freely then you will need to adjust the balance cock. There are two ways this can be done. If only a small amount of adjustment is required, it is possible to raise the balance cock by placing a small piece of paper or tin foil under its foot. If this does not cure the problem you may need to bend the balance cock up. Only do this with a pair of brass-faced pliers and be very careful only to make small adjustments and test each time whether you have corrected the fault.

- The lever pallets may unlock incorrectly. As the balance moves backwards and forwards the impulse pin moves the lever first one way then the other. It is stopped from moving too far by the banking pins (one either side of the lever). The position of these is critical to the correct operation of the watch. If they are too close to the lever, the pallets will not unlock from the escape-wheel teeth. If they are too far apart, the pallets may interfere with the escape wheel. The impulse pin may also be unable to engage with the lever fork so preventing energy being transferred to the balance.

  With going barrel levers the banking pins are usually formed from a screw that passes through the top plate on which is mounted the banking pin, offset from the centre. Therefore the position of the banking pin relative to the lever can be adjusted by twisting them with a screwdriver. On older watches they are usually formed of brass pins inserted through the bottom plate.

  Always adjust the banking pins with the balance removed. With the movement wound up, move the lever from one side to the other with a piece of peg wood; it should snap over and the escape-wheel tooth should disengage. It should rotate a small amount and the opposite lever pallet should engage with the next tooth on the escape wheel. Flipping the lever back the other way should repeat the cycle. In all cases the pins need to be adjusted so that the lever pallet just unlocks from the escape-wheel tooth.

- Sluggish balance may be due to a weak main spring. This is particularly true of verges. First check that the balance swings freely with the spring fully unwound. If it is in order, try replacing the main spring with a stronger one. Be careful not to fit one that is too strong, as this may damage the movement. Before changing the spring, make sure that you have checked for free running of the train; the problem could be binding pivots on the wheels etc.

Damaged pivot holes

To close a hole, place the plate on the staking tool with the inside surface facing up and, with a round-ended stake, lightly tap around the pivot hole (being careful not to hit the hole itself as this may make it larger) until it is a little smaller than required for the pivot. Then using a suitably sized burnisher, open the hole so that the pivot just fits without friction. This is something that can only really be judged by trial and error, so take it slowly.

If the hole is too large to be closed up by staking, a new bush may need to be inserted. This consists of a small brass disc with a hole in the middle. The old pivot must be drilled out and then a reamer used from the inside to open the hole until the bush is just too big for the hole. The reamer should have produced a slightly conical hole in the plate. The bush can then be fitted in place by positioning it over the hole (from the inside) and, using a flat-faced stake, tapped into place. Once you are happy that it is correctly positioned, using a burnisher open up the hole as above.

Damaged jewels

It is quite common to find that one or more of the jewelled pivot holes have become damaged. This means that the pivot can no longer run smoothly. There is not much that can be done with a damaged jewel but to replace it. This is fairly straightforward if the jewel is mounted in a removable metal holder and held in place with screws. Remove the screws and, using a piece of peg wood, push the damaged jewel housing out. Then from a scrap watch or stock, fit a suitably sized replacement. Not only does this need to fit the hole in the plate but the pivot hole needs to be the correct size. Once fitted, secure it in place with screws.

If the jewel is fitted directly into the plate (known as rubbed in), then it is more difficult to replace. First push out the old damaged jewel with peg wood, and then carefully open the hole with a suitable stake. You then need to select a new jewel (these can be obtained from speciality suppliers). To fix the new jewel in place, you need to close the hole around it. This is done using a round-ended staking tool, running it round the depression and applying a little pressure. This should cause the metal to fold over the edge of the jewel; be careful not to apply too much pressure as this could crack the jewel.

A bent pivot

Sometimes the pivot can be bent; it needs to be straightened before the watch will work properly. Great care is needed to prevent the pivot from breaking off. You need to use a pair of pliers with brass-faced jaws. The ends of the jaws should be gently warmed over a spirit lamp or gas stove flame. Holding the wheel between the thumb and finger, close the jaws of the pliers over the bent pivot (do not close too firmly) and at the same time pull away from the pivot. This action may need to be repeated a few times until the pivot is straight. No matter how careful you are, in some cases the pivot will break off. For this reason, it might be worth trying this technique on scrap wheels before you try it on a wheel from a good watch.

A broken pivot

If you find a wheel with a broken pivot, there are a number of ways it can be fixed.

- One way is to use a replacement wheel from a scrap watch. For this to be successful, the wheel diameter, pinion and number of teeth must be identical; the distance between the pivots must also be of the same length or it will not fit between the plates. In practice this probably means that the part must come from a watch by the same manufacturer. Note that it is much more difficult to interchange parts from watches made before the late nineteenth century. This is due to the fact that these were in part made or finished by hand, and therefore exhibit greater variation from one watch to the next. With later watches (particularly Swiss and American ones), parts are much more likely to be inter-changeable within a particular manufacturers.
- The second method is to remove the pinion from the wheel and replace it with a good one. Again the pinion must be of the same size and number of teeth. The pinion is fixed to the wheel by a friction fit. To remove it, place the wheel on a staking tool with the pinion through a hole a little larger than its diameter. Then using a suitable staking tool, tap the pinion until it drops away from the wheel. The new pinion is fitted in place by turning the wheel over and placing it above a small hole (just bigger than the diameter of the new pinion) and using a suitable stake; lightly tap it until it is flush with the wheel.
- A third way is first to remove the wheel from the pinion (as detailed above), then heat the broken end in a spirit lamp or gas stove flame until it glows a dull red colour. This softens the steel making it possible to drill into the end. Care must be taken not to overheat the part. The next bit is quite tricky, as you need to drill a small hole about 2–3mm deep into the end where the broken pivot used to be. This hole is used to fit a new pivot taken from a scrap one. Note that there are special tools that are designed to help with drilling

the hole (called a watchmaker's pivoting tool). The diameter of the hole is dependent on the diameter of the replacement pivot, and should be sized to give a tight fit. The new pivot is then tapped into the hole using a suitable stake. If the hole is a little too large, the pivot can be held in place with a little superglue.

A broken spring

If the spring is damaged in some way, it will need to be replaced. The replacement can come from old stock (obtainable from auctions etc.), you can reuse a spring from another watch of similar design or you can buy a new one (available from watch material suppliers), although this is really only true for going barrel levers. It might be a good idea, if you intend to repair several watches, build up a stock of parts and old broken watches to help with repairs.

To remove an old spring, first remove the spring barrel cover and arbor, then carefully lift up the inner end with a screwdriver while keeping the rest of the spring in place with your finger. Be aware that a spring can do a lot of damage if it is not removed with care. Pay particular attention to protecting your eyes. It would be a good idea to invest in a pair of safety glasses (available from most DIY shops). Working with the open end of the spring barrel pointing away from you, slowly allow the spring to unwind from the barrel. Make sure you hold onto the barrel firmly or it may fly off as you come to the end of the spring. To fit a new one, fix the outer end of the spring to the inside of the spring barrel (make sure it is the right way round), then push it into place while rotating the barrel, until you reach the end. Fit the arbor and make sure that the hook engages with the end of the spring; if it does not, carefully bend the end in with a pair of pliers. Apply a small amount of watch oil and refit the cover.

If the spring has broken near the inner end, it is sometimes possible to repair it. Remove it as above and take the broken end and heat approximately 1–2cm in a flame until it is red hot, then let it cool naturally, which will soften the steel. Hold the softened end over a hole on the staking tool and place a pointed stake over it. Tap the stake a couple of times so that it forms a dent in the spring, then use a file to remove the bump, which should leave a small hole. Using a reamer, open the hole up until it is the desired size (i.e. it fits over the hook on the spring barrel or arbor depending on which end has broken).

The outer end of a spring can be held by a number of methods: a hook, a shaped piece of metal riveted to the end which fits into a slot in the spring barrel, or with the end bent over and held by a groove on the inside of the barrel.

If the spring is held by a hook, proceed as for the inner end described above. If it is held by a shaped piece of metal, you can try and remove the original piece and, after punching a hole in the spring, re-rivet it to the end. This may not work as there might not be enough metal to make a good joint. If this is the case you can make a new one by filing a piece of steel to shape and rivet this to the end of the spring. This is quite difficult, and an easier way might be to convert the spring barrel to a hook fixing. Drill a hole in the side of the barrel, on the centre line; try to angle the hole a little so, when the hook is inserted, the spring will not slip off. Using a tap, cut a thread into the hole. Take a screw that will fit the cut thread and screw it in so that enough of the end protrudes into the barrel to provide a good hook for the spring. Cut off the screw on the outside and then file the stump flush with the barrel.

If the spring end is held in place via a groove cut in the barrel, you need to create a new hook on the end of the spring. Heat about 1.5cm of the spring to soften the steel, and then bend it over so that it doubles back on itself. Then break a piece off an old spring about 1cm long using a pair of pliers to snap it. Fix this in place in the end of the spring by squeezing it with pliers.

Spring barrel problems

In most cases a spring barrel consists of a body, into which the spring is fitted, and a lid with the spring arbor held between them. In going barrel watches, the body usually has teeth around the edge, which engage with the centre wheel. With fusee movements, the body of the spring barrel is smooth. In both cases, the lid is held in place by a groove around the top of the body into which it fits tightly. As can be seen from the drawing below, the inside edge slopes in (greatly exaggerated in the drawing), and can become worn by repeated fitting of the lid. This will lead to the lid popping off during winding. To fix this problem, you will need to tap the outside edge lightly all the way round, with a small hammer. This should cause the top edge to lean in and hence make a tighter fit with the lid.

A spring barrel lid grove

Fusee problems

Sometimes as the watch runs down and the fusee chain winds on to the spring barrel, the chain can catch on the fusee hook and not lie flat on the spring barrel. This can usually be fixed by putting in a new hook slot. Remove the spring barrel from the movement and then take the spring out. Using a small drill (0.2–0.3mm), drill a hole a little above and to one side of the original slot. You need to angle this hole into the spring barrel as shown below. Then, using a reamer, make it the right size for the fusee chain hook. Reassemble the movement and check that this has cured the problem. Make sure that you do not put the new slot too high as this could mean that the chain will interfere with the underside of the bottom plate.

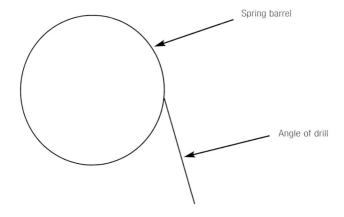

Spring barrel

Angle of drill

Drilling a new hook slot

A broken fusee chain

If the fusee chain is broken but you have both parts (and both hooks are in order), it is sometimes possible to joint them. A fusee chain consists of a number of steel links riveted with steel pins, much like a miniature bicycle chain. There are two ways of proceeding. The first is to take one piece of the broken chain and, using a craft knife, carefully lever away the link on one side. Then turn it round and lever off the broken piece. This should leave the two outer parts of the link connected to the rest of the chain (see the left-hand drawing below). Now take the other piece and, using the craft knife, remove the outer two pieces to leave behind a central link which should still have the original pin in place (see the right-hand drawing below). If you fail to achieve the correct arrangement, repeat the procedure (be careful not to take too much off the length of the chain, as it may become too short for the watch).

163

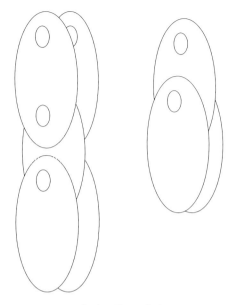

Broken fusee chain

Now bring the two parts of the chain together, making sure that the hooks face the same way (both pointing out or both pointing in). Using tweezers, push the outer pair of links over the inner one so the holes in the outer links go over the pin. When you are happy that they are positioned correctly, press the outer links onto the inner piece with a pair of pliers. Then place the chain on the surface of a staking tool or other hard, flat, steel surface and tap a few times with a hammer. Turn it over and repeat. Check the joint with a x10 loupe; it should look smooth, with the ends of the pin filling the hole. Pull the chain to make sure the link is strong.

The second method is similar, but you should remove the pins completely from both ends, then take a piece of steel wire with a small diameter (a dressmaker's pin heated in a flame to soften it and filed to fit is ideal). Hold both pieces of the chain together so that the holes line up, and push the wire through all three parts. Place the assembly over a suitably sized hole on a staking tool and lightly tap the pin so that it is a tight fit. Cut off the wire on both sides, as close to the chain as possible. Then using an Arkansas stone, remove most of the pin, leaving only a small amount (approximately 0.25mm) on either side. Place the chain on a flat, hard, steel surface and tap the top of the pin until it is flattened into the link, turn it over and repeat on the other side. Smooth it off flush with an Arkansas stone. Pull the chain to check its strength.

If the watch is not to be much used, it is possible to substitute a brass pin for the steel one but it will not last as long.

The above two processes can also be used on lengths of scrap fusee chains, so never throw these away. Always make sure that both pieces of chain are of the same size.

A broken fusee chain fixing pin

Another common fault with a fusee is that the small metal pin that the fusee chain hooks over can be pulled away from the fusee. To repair this, if the original holes (top and bottom) are fine, pick a suitably sized tapped steel pin (it is possible to use a brass pin but it will not last as long), push it from the underside through the two holes and tap it lightly. Cut off both ends as close as possible to the fusee and then, using an Arkansas stone, smooth each end until it is flush with the surrounding metal.

A damaged balance

There are several problems that affect the balance. These include the following:

- A damaged hair spring
- Bent pivots
- Broken pivots
- A bent or misaligned balance
- A broken impulse pin

A damaged hair spring

Because the hair spring is very fine, it is easily damaged, so whenever you are handling the balance, be very careful. If the spring has become distorted, it can sometimes be repaired by using a pair of tweezers and carefully bending it back into shape. If the problem is that the spring is no longer circular or some of the loops are out of alignment, then it can be realigned using two pairs of tweezers. First the spring needs to be removed from the balance. Using a craft knife, lever up the spring collet by getting the knife right underneath it. Be very careful that the spring does not fly off. Place it on a piece of white paper and then hold it just before the distorted section and push or pull it until it is realigned. Continually let go of the spring completely to check whether it is fixed. Repeat until the spring is circular and none of the loops overlaps.

Choosing a replacement hair spring is quite tricky. Ideally the replacement should be the same size and apply the same rotation force as the original spring. The force the spring applies is one of the things that affects the rate of oscillation of the balance. Therefore if this does not match the original, the watch may run faster or slower than it should. There are specially designed devices that can measure the rate of oscillation of a balance but, as the original hair spring will probably be damaged, this is of little use. There are methods that can be used to calculate the required spring but this will probably involve more information than is available. The simplest way is to choose a spring that is of a similar size to the original and fit it to the balance, reassemble the movement and test it for accuracy.

Bent pivots

These can be dealt with in the same way that the wheel pivots are repaired (page 158).

Broken pivots

These can sometimes be repaired in the same way as broken wheel pivots (pages 159–60). Alternatively it might be better to replace the whole balance staff. Old stock of staffs can often be bought from watch repair suppliers and on-line auctions. To replace a staff, carefully remove the hair spring as detailed above (if it is above the balance) then, using a pair of brass-faced pliers, remove the roller and impulse pin assembly by twisting it and at the same time pulling it away from the balance. Great care must be taken or you can easily break the impulse pin or distort the roller. If the balance spring is under the balance, then you will need to remove the roller assembly first. Having done this, place the balance on a staking tool over a hole large enough to allow the balance staff to fall through and, using a suitably sized stake, tap the staff until it drops away from the balance. You then need to select a new staff of the same size. This then should be tapped back into the balance. Note that not only does the staff need to be the same overall length as the original one, but also the length of each section (above and below the balance) needs to be the same or there is a danger of the balance catching on other parts when in place.

Before refitting the hair spring and impulse plate, try the balance to make sure that it fits. It should spin freely (rotate for a few seconds) with the balance cock screwed in place. If the balance does not turn freely, either the staff is too long or the pivots are too big for the pivot holes. If the staff is too long, you have two choices: you can either replace it with a slightly shorter one or lift the balance cock by a small amount. A common trick in this situation is to place a small piece of paper, cut to size, between the main plate and the underside of the balance cock.

If the pivots are too large for the pivot holes, this can be rectified using a Jacot tool, which is not unlike a small hand-driven lathe that has been specially adapted to adjust the size of pivots. Use the gauge that comes with the tool to establish the correct size for the pivot by measuring the old balance staff. If both ends are broken, you will need to try the pivot at each stage until it fits. Select a runner that is just smaller than the current size of the pivot, do not go straight to the final size as this will put too much stress on the pivot and it may break off. Place the balance on the tool and adjust its position by moving the runner bar in and out until the pivot just rests in the runner, and then lock it in position. Wind the knob in until the drive pin engages with the balance arms and lightly hold an Arkansas stone over the pivot resting on the tool, then using the bow slowly rotate the balance keeping light pressure on the stone. Do this until the balance pivot cannot be reduced any further. Lift the balance away and unlock the runner bar, rotate this to the next position and set up as before. Repeat this until the pivot is the same size as the original pivots. Repeat for the other end, then try it in the watch and check that it spins freely.

Assuming that it is spinning freely, reassemble the rest of the balance. Place the balance on a staking tool with the bottom pivot uppermost (you will need to use a stand-off to prevent the balance from becoming distorted). Place the impulse roller over the balance staff and tap it home with a thin stake (make sure that the stake is thin enough not to damage the impulse pin). Turn the balance over and place it on the staking table so that the balance staff and impulse roller drop through a large enough hole; the balance should be flush with the surface of the table. Place the hair spring over the upper pivot and tap it into place with a stake. Make sure that the spring and impulse roller are aligned as they were before it was dismantled. Reattach the hair spring stud to the balance cock and then fit the complete assembly on the main plate. If all has gone well, the balance should oscillate freely for a few seconds when shaken.

If the hair spring is mounted under the balance, it needs to be fitted before the roller assembly.

A bent or misaligned balance

Sometimes, the balance can become bent or distorted due to heavy handling (this is particularly true of cut balances). It is fairly easy to realign the balance. Hold it with a pair of pliers and pull or push the bent section back into place. You are aiming to return the balance to as near a perfect circle as possible, while also keeping it flat in the horizontal plane.

A broken impulse pin

A common problem on lever watches is the impulse pin breaking off. It can be replaced from old stock. It is usually held in place with shellac. Carefully remove the impulse roller as described above (noting the position on the balance staff). Using a sharp metal point, remove the remains of the old pin and shellac (methylated spirits will remove the shellac). Choose a new impulse pin of the same size and check that it fits the hole correctly; also check that it is the correct size for the lever fork. When you are happy that it is the right size, fit it in place, making sure that the flat faces forward, and glue it in place with a small drop of superglue on the top. Before the glue dries, check that the impulse pin is vertical to the table. Replace the impulse roller as above (making sure it is in the same position as before).

Dial repair

As mentioned earlier, dials can often be renovated by soaking in a bathroom cleaner. This will deal with surface grime and dirt trapped in a hairline crack, and apart from this there is not much else that can be done with regard to cracks. Chips can be filled using epoxy-resin glue. There is a kit available from Walsh that is designed to fix common problems with enamel dials (see appendix A). This contains a clear epoxy resin and coloured pigments so you can match the repair to the original enamel. There is also ink to redraw the numerals. However, to make a good job will require both patience and practice.

Another way to repair damage is to use a bath-enamel repair kit. This works fairly well, but it often is not possible to match the colour.

Replacement hands

If you are lucky you will be able to reuse the existing hands, but quite often they will be either missing or damaged and therefore you will need to fit replacement ones. As with most watch parts it is worth keeping an eye out for watch hands and purchasing them for spares.

The hour hand is held in place by a friction fit over the end of the hour wheel. A replacement must have a similar-sized hole and it should also be the correct length (the point should just come to the inner edge of the numerals). It is possible to change the size of the hole by a small amount. To increase it, place the hand on the staking tool over a larger hole and place a long tapered stake through the ring on the hand and then tap the stake lightly. Be careful not to hit it too hard or you might split the ring. To make the hole smaller, place the hand on a staking tool (on an area without any holes) with the front facing down. Place a staking tool with a cone slightly larger than the rear tube over it and lightly tap it (see drawing). The effect should be to close the tube enough to make a tight fit on the hour wheel.

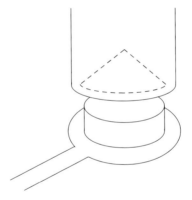

Closing the tube on an hour hand

The minute hand is slightly more complex. There are two methods of fixing them. On most fusee watches, the hand is fixed by a square which forms a friction fit over the end of the cannon pinion. The replacement needs to be close in size to the original. With gold hands, if the square is too large it can be made smaller by placing it face down on a staking tool and, using a flat-faced stake, tapping it a few times. This has the effect of spreading the metal and therefore making the square hole smaller. If the square hole is smaller than the cannon pinion, it can be made larger by filing it; hold the hand in a pin vice and use a needle file with a square profile, but be careful that you hold the end of the hand near the hole and not the hand itself.

The other type of minute hand has a round hole and is held in place by a friction fit on the end of the cannon pinion. If the hands are gold, the hole can be closed in the same way as a hand with a square fitting. If it is too small, the hole can be opened with a reamer.

These techniques can be tried with steel hands, but due to the hardness of the steel, it may be difficult to carry them out successfully.

The second hand is held in place by a tube fitted over the extended pinion from the fourth wheel. To fit a new one you may need to open the tube with a reamer in order to get it to fit. If the tube is too large, it is possible to squeeze it with pliers (do not squeeze too hard as it is easy to collapse the tube).

Remember that the hands are very important to the 'look' of the watch and it is therefore important that you fit the appropriate style and size. If the original ones are missing, check in books for watches of a similar age to gauge which type would best suit.

A loose or slipping cannon pinion

If the watch appears to run well, but keeps very poor time, or the hands stop moving altogether, it is often due to a loose or slipping cannon pinion. Remove the hands, dial and motion work. Check the cannon pinion; it should be able to rotate on the centre wheel arbor, but there should be some resistance. If it is too loose then the centre wheel cannot turn the cannon pinion or the hands. To fix this you need to make the hole in the cannon pinion a little smaller. If you have a staking tool, place the cannon pinion on a flat stake and, using a round-ended stake, tap it a few times near the centre while rotating it. This should close the hole enough to make the fit on the centre wheel arbor tight (but it must still be able to be twisted so that the hands can be set).

Alternatively it may be possible to fix the problem by putting a small piece of aluminium foil between the arbor and the inside of the cannon pinion.

Loose pillars

The pillars that hold the top and bottom plates together are usually riveted into the top plate. On dismantling the movement, it may be found that some of these are loose. To fix them put the end of the pillar into a hole on a staking tool table, just bigger than the top of the pillar but small enough for the flange to rest on the top of the staking table. Using a round-ended stake, lightly tap the centre of the pillar where it fixes to the top plate, until the pillar is firm.

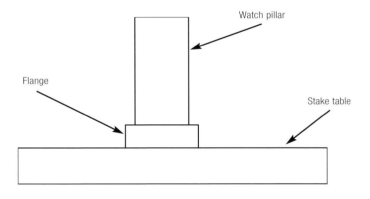

Tightening pillar on top plate

This process can be applied to the dial plate as well.

Listening to the watch

A lot can be gained by listening to a ticking watch. If it is a lever movement, it should have a clear tick (caused by the lever hitting the banking pins); sometimes there may be a slight metallic 'ting'. If the sound is not a clear tick, it may indicate that something is not quite right. After listening to a few watches you will get an idea of what sounds good and when there is a problem. The sort of thing to listen out for are the sound of the hair spring rubbing on the balance (a slight rasping sound), or the edge of the balance catching on the underside of the balance cock.

With a verge watch, it is also possible to diagnose problems by listening to it ticking. Because of the design of the verge, it has a very different sound from a lever. This is due to the fact that the balance is not detached (the crown wheel is in contact with the pallets at all times). This tends to give a deader sound to the ticking. One thing to listen to is whether the tick is even; if not it may indicate that the balance is not correctly aligned – the pallets are not evenly spaced either side of the crown wheel axis with no power applied from the spring.

Fob and wrist watches

Having gained experience on pocket watches, it should be fairly straightforward to transfer these skills to fob and wrist watches, which often have the same type of movements as pocket watches, but on a smaller scale.

When to give up!

No matter how experienced you become, there will still be times when a particular problem defeats you. When this happens, the best thing to do is to put the watch to one side (make sure you keep all the bits together and record the work you have already carried out) and try again in a few weeks' time. You will be surprised how an apparently insoluble problem suddenly becomes straightforward. This may be due to the fact that your skills have improved or that a fresh look at the problem has given you a new insight. Remember, watch repairing should be a hobby, not an obsession!

TOOL MAINTENANCE

I t is good practice to maintain all your tools in top condition. This makes them easier to use and less likely to cause damage. If any become broken or worn, they should be replaced or repaired as detailed below.

Screwdrivers

Always maintain a good sharp blade, as this will prevent damage to the screw head. Use a screwdriver blade sharpener and stone; fit the screwdriver in the holder, check that the blade forms the same angle with the stone as it is currently and that it is at right angles to the stone. Run the sharpener backwards and forwards a few times and then check the blade with a loupe. When you are happy, turn it over and repeat on the other side. Finally use a stone across the end of the blade until it is flat and square.

Tweezers

For these to work properly, their internal faces need to be flat and parallel. Use an Arkansas stone to smooth the internal faces. Also run it over the points to keep their shape.

Pliers

Again the internal faces need to be smooth and parallel. Use an Arkansas stone to smooth them.

WATCH DATING

I f the movement and case have always been together, it is fairly straightforward to date a gold or silver watch from the hallmarks. Be careful when checking the date letter against the lists (see appendix A) as it is all too easy to pick the wrong series and date the watch too early or too late. Try to consider the style of the movement as well. If it is signed, check in a reference book (see appendix B) for dates of when they were working. This, coupled with the case style, should give a good idea of the date of the watch.

With non-hallmarked cases (gilt, nickel etc.) or continental ones (which tend not to carry any date), you really only have the watch movement and case style and possibly a signature or manufacturer to go on.

Watches dating from the seventeenth and eighteenth centuries were often housed in pair or even triple cases. This was partly to protect the watch and partly to make it look more important. The outer case was usually very heavily decorated or covered in tortoise shell or shagreen (shark skin). These cases were usually made of gold, silver or gilt, sometimes called mercurial gilding.

As the eighteenth century drew to a close, the cases became simpler, but gold, silver and gilt cases were still the most common. By the nineteenth century cases were usually plain or with geometric designs (engine turning), and towards the end of the nineteenth century, it became common to find watches housed in steel or nickel cases. Note that the silver pair case continued in use well into the nineteenth century.

Towards the end of the nineteenth century and into the twentieth century, more and more mass-produced watches were made, which were often housed in plain silver, gold-plated (sometimes called rolled gold or filled gold), steel, nickel and chrome-plated cases as well as, but to a lesser extent, gold and silver.

From the earliest days of pocket watches, they have always had some sort of dial to indicate the hours and later minutes and seconds. In the early days, these were made of similar materials to the case, i.e. gold, silver or gilt brass. Later, during the seventeenth century, enamel dials became increasingly popular. These were made by fusing ground glass coloured with pigments onto a copper disc and are sometimes incorrectly referred to as porcelain dials. In the early days, they were not usually white, but were either cream or had a blue tinge. Up until the latter part of the eighteenth century, they tended to be convex, but by the beginning of the nineteenth century they were becoming flat and their colour was now more commonly white. Also, with the increasing accuracy of watches at the end of the eighteenth century, dials started to be made with a subsidiary seconds dial; these often became sunken in the latter part of the nineteenth century.

One thing to look out for is recasing. This may have been done for legitimate reasons, such as the original case becoming worn out. The new case will bear the hallmarks for when it was manufactured and this will be at variance with the true age of the movement. The other way a watch movement may end up in a case of a different age is where a seller or collector has found a suitably sized case and fitted the movement into it with the deliberate intention of passing it off as a good watch (known as a 'marriage'). One of the things to look out for is whether the serial numbers match. On most English watch movements you should find a serial number and, if there is one on the case, they should match. The absence of a serial number on the case is not necessarily a problem. Check for the signs, such as whether the winding hole lines up with the winding square, whether the movement hinge fits, whether there are modifications to the case, and finally, whether it looks right. With experience, you will develop a 'feel' for this.

WATCH SIGNATURES

Quite a lot of the watches you may come across will have a maker's name engraved on them. Most books refer to these signatures as being the watchmaker. This is not strictly true, as in all but a very few cases, as far as nineteenth- and twentieth-century watches are concerned, the signature is that of the retailer or, at best, the watch finisher. Most of the watches will have been made by outworkers. These were individuals or families working from their own houses (later in factories), each making a particular part or parts. These were then collected together and assembled into what were known in the trade as roughs, before being sold on to the finisher or retailer, who would complete the watch and add their signature to the movement. Even in the eighteenth century it was usual to follow this process. For more details on the watch trade see *Watch-making in England 1760–1820* by Leonard Weiss.

SELLING POCKET WATCHES

You may be a collector, in which case the need to repair a watch is an end in itself. However you may want to consider selling watches, either as a way of making money or to subsidize your collecting. Any watch that can be described as working, and by that I mean able to run with reasonable accuracy for at least twenty-four hours between winding, will always fetch a higher price. You may wish to sell to an antique dealer or shop, but be aware that the price you receive will be a lot lower than the retail sale price that you see in the shop window. This is because the retailer has to cover their overheads and be able to make a profit. They may also have to carry stock for quite a long time before it sells.

An alternative to selling via a shop is to use some form of auction. There are two types: the more traditional auction house and the newer and increasingly popular on-line auction. With the traditional type, you give the watch to the auction house and they are responsible for describing the item, printing the catalogue and collecting the money from the buyer. For this they charge you, and with the larger companies also the buyer, a fee based on a percentage of the final sale price. Unless it is a fairly valuable watch, it is probably not worth considering this route. For lower-value watches (and even some more valuable ones), it might be better to use an on-line auction site. The best known is probably eBay, but there are others that deal with pocket watches. They all operate in much the same way. You are responsible for listing the item, collecting payment and posting the item to the purchaser. You will be charged a fee for both listing the item (even if it does not sell) and a final value fee, based on a percentage of the sale price. There are other costs depending on the payment method you choose. The advantage of an on-line auction is that you are in full control: you set the start price, how it is described etc. To get the best price you need to provide a good, detailed description and high-quality photographs. These need to show more than one view, including the movement. Try to be honest about the watch; make sure you mention faults as well as the good points. Remember a satisfied customer will come back but a dissatisfied one will tell everybody to stay away from you.

Your potential market is worldwide and there are a large number of very knowledgeable collectors who will pick up on a good or unusual watch.

It is best not to start by selling valuable watches until you have built up a reputation. Once you are established, you may consider selling more valuable items.

CONCLUSION

I hope this guide has been of use to you in developing a new and enjoyable hobby or maybe a business opportunity.

If, having read this guide, you have any comments or suggestions, I would be more than happy to hear them. Please email them to: Westhill.watches@btinternet.com

I cannot promise to answer all enquiries but I will do my best.

APPENDIX A: Sources of information on the internet

www.agthomas.co.uk Horological and jewellery supplies.

www.antique-pocket-watch.com Useful site, especially for information on American pocket watches (Waltham, Elgin etc.).

www.bhi.co.uk The British Horological Institute, a very good source of technical information plus links to other useful sites.

www.ebay.co.uk Good site to buy (and sell) pocket watches plus tools and materials.

www.horologia.co.uk This site has very good illustrations of a nineteenth-century English lever pocket watch plus an extensive glossary and lots of other useful information, including hallmarks.

www.horologysource.com The Horology Source, a very useful site with a wealth of information, details of escapements and mechanisms including animated drawings, links to other sites etc.

www.hswalsh.com Watch materials plus a very useful dial repair kit.

www.rnhorological.co.uk A good source of tools and watch keys (fast delivery).

www.shentonbooks.com A good source of books on horology, with fast and friendly service.

www.tickintimeworldofwatchtools.co.uk A good range of tools.

www.watchtool.co.uk A supplier of Bergeon tools and materials.

www.925-1000.com An on-line encyclopaedia of silver hallmarks from around the world, with a very good section on UK marks.

APPENDIX B: Further reading

Britten, F.J. *Britten's Old Clocks and Watches and Their Makers* (Pub Marking Enterprises, 1986) (first edition 1899). This book contains a great deal of useful information, including a list of clock and watchmakers. It has been revised and reprinted many times.

Britten's Watch and Clock Maker's Handbook, Dictionary and Guide (Bloomsbury Books, 1987). Contains lots of useful information on watches, with good illustrations.

Chamberlain, P. *It's About Time* (Holland Press, 1978) (first edition 1941). Deals with the more unusual movements and has a very good biographical section covering early watchmakers.

Cutmore, M. *The Pocket Watch Handbook* (David & Charles, 1985). This is a good introduction to pocket watches, and covers the history from the first watches in the sixteenth century right through to the low-cost everyday watches of the late nineteenth and early twentieth centuries, plus notes on repair, collecting etc.

Cutmore, M. *Watches 1850–1980* (David & Charles, 1989). This book covers the more everyday pocket watches that are readily available to the ordinary collector.

Daniels, George *Watchmaking* (Philip Wilson, 1981). George Daniels designs and makes pocket watches; he manufactures every part, including the dial and case. This book covers the whole process from start to finish and is therefore a good source of information if you need to replace any part yourself or just want to understand how a pocket watch is made.

de Carle, Donald *Complicated Watches and Their Repair* (NAG Press, 1956)

de Carle, Donald *Practical Watch Adjusting and Springing* (NAG Press, 1964)

de Carle, Donald *Practical Watch Repairing* (NAG Press, 1969)

de Carle, Donald *Clock and Watch Repairing* (Hale, 1981)

Gazeley, W.J. *Clock and Watch Escapements* (Hale, 1992)

Gazeley, W.J. *Watch and Clock Making and Repairing* (Hale, 1993)

Kemp, Dr Robert *The Fusee Lever Watch* (John Sherratt & Son, 1981). This book covers in detail the English fusee lever pocket watch, including notes of some of the many variations (Rack, Massey etc.).

Loomes, Brian *Watchmakers and Clockmakers of the World* (NAG Press, 2006). This contains the details of thousands of clock and watchmakers, listing their dates etc.

Priestley, Philip T. *Watch Case Makers of England* (National Association of Watch and Clock Collectors, 1994). This is an American publication which includes a wealth of information regarding watch case manufacture in England, plus a list of makers in the three main areas of production (London, Birmingham and Chester).

Weiss, Leonard *Watch-making in England 1760–1820* (Hale, 1982). A very well written and fascinating account of the watch trade in Georgian England, detailing all the operations needed to produce a finished watch.

Whiten, Anthony J. *Repairing Old Clocks and Watches* (NAG Press, 1979). Well illustrated, very clear line drawings, covers watches (pocket and wrist) and clocks.

APPENDIX C: Some useful terms

- **Balance cock** – The upper pivot of the balance which either partly or fully covers it. In early watches, it was heavily pierced and engraved, during the nineteenth century it became smaller (only partly covering the balance). Usually only engraved, sometimes they are completely plain.
- **Chronograph** – A description that is applied to many types of watches, usually indicating that the watch has a large centre second hand. The watch often has some sort of stop work, whereby the train can be stopped and started. A variation is the fly-back chronometer, where the centre second hand and a subsidiary minute hand are controlled separately from the main train and can be started, stopped and returned to zero without upsetting the running of the watch.
- **Cylinder** – A type of escapement designed by George Graham in about 1725. Later to become the most common type employed by Swiss and French makers during the nineteenth and twentieth centuries.
- **Detached lever** – A type of escapement designed by Thomas Mudge in 1757, in which impulse to the balance is only given for a small fraction of the total rotation of the balance. This reduces the friction on the balance and hence improves the accuracy of the watch. The most common form on English watches of the nineteenth century is the table roller lever.
- **Fob watch** – Usually used to describe a small (under 35–40mm in diameter) pocket watch, quite often intended for use by women.
- **Fusee** – A device to regulate the uneven force of the main driv-

ing spring as it runs down. It consists of a spiral groove cut into the side of a cone. This is linked via the fusee chain to the spring barrel. Because of the varying diameter of the fusee, the force acting on the watch train is balanced against the weakening force from the spring as it runs down.

- **Fusee chain** – Flexible steel chain (not unlike a miniature bicycle chain) used to link the spring barrel to the fusee.
- **Gilding** – A thin layer of gold applied to a base metal, in watches usually brass. This was originally done by applying an amalgam of gold and mercury to the metal and then heating it in an open fire, driving off the mercury as a vapour and leaving a thin deposit of gold. The process is not to be recommended, as mercury vapour is highly poisonous. During the nineteenth century this process was replaced by other methods such as gold plating (where gold is deposited using an electrical current), rolled gold or gold filled (where a layer of gold is laminated over a base metal, usually brass).
- **Going barrel** – A watch in which the spring barrel drives directly onto the train, made possible when improvements in the manufacture of steel made it possible to do away with the fusee.
- **Hunter case** – A design of watch case where the whole of the dial is covered by a metal cover which flips open when the pendant button or winding crown is pressed. A variation on this is the half-hunter where the front cover has a small opening (less than the diameter of the dial) through which the hands can be viewed without opening the cover. The outside of the front of the cover is usually marked with the hours as well.
- **Jewelling** – A system by which the pivots of the wheels (cogs) are made to run in jewels which have had small holes drilled through them. This arrangement reduces friction and wear so that the operation of the watch is usually more reliable.
- **Keyless** – An arrangement that allows the watch to be wound up via a winding crown protruding from the top of the watch. By pulling the crown out, the hands can also be set. A variation on this is known as 'pin set' where a small pin near the crown is pressed in to allow the hands to be set.

- **Loupe** – A high-quality magnifying eyepiece, usually with a high magnification (x10 or more).
- **Maintaining power** – A device fitted to the fusee to provide power to the train during winding. In order to wind a watch, force must be applied in the opposite direction to the running of the watch, which can cause the watch to slow down or stop.
- **Marine chronometer** – A very accurate type of watch or clock used to measure time at sea with the purpose of determining longitude. John Harrison designed the first truly accurate watch for this purpose in the 1760s. Most marine chronometers use some variation of the spring-detent escapement (see page 18 for details).
- **Motion work** – An arrangement of wheels (cogs) under the dial that drives the hour and minute hands.
- **Overwound** – A common diagnosis applied to a non-working watch. In most cases it is incorrect. If you apply too much force in winding the watch, the spring will 'slip' or break or, in the case of a fusee, the chain will snap. It is usually used to describe a watch that has been fully wound but does not tick. The spring is therefore fine, and the reason the watch will not work is due to another problem.
- **Pallet** – A metal or jewel that protrudes from another part (balance staff etc.) against which the parts of a watch mechanism engage.
- **Pendant** – The protuberance on top of a watch case usually carrying a ring or bow to allow the watch to be fixed to a chain. Later this became the winding crown in keyless watches.
- **Regulator** – A device that allows the rate (accuracy) of the watch to be adjusted by changing the effective length of the balance spring (hair spring). The longer it is, the slower the rate of the watch, and the shorter it is, the faster the watch will go. There are several ways in which this can be done. In old verge watches it is under the balance. On later watches this often moved to the balance cock. It became increasingly complex, particularly in early twentieth-century American watches.
- **Repeater** – A watch where the hours, quarters – and sometimes minutes – are struck on gongs or rods, much the same as a

grandfather clock. It usually has a button that can be pressed to repeat the chimes, hence the name repeater.

- **Verge** – The first escapement to be used in watches. Known as a frictional rest type of escapement because the drive is constantly in contact with the balance.

INDEX